Invisible Visitors From Outer Space

Cosmic Rays

The Story from the Beginning to the Present

George R Steber

Invisible Visitors

From Outer Space

Cosmic Rays

The Story from the Beginning to the Present

This book is humbly dedicated to the memory of my friend and mentor, Professor G. L. Elmergreen who inspired my interest in engineering, science and mathematics during my formative years at the University of Wisconsin in Milwaukee. I also wish to thank my parents for their unconditional love and giving me the unfettered freedom to choose my life's work.

Above all, I wish to thank my dearest companion Professor Lynne Reynolds for coaxing me to write this book, especially when I had all but given up. Her love and encouragement were an inspiration to me. She is a true sweetheart, if there ever was one!

Everything is determined ... by forces over which we have no control. It is determined for the insect as well as the star. Human beings, vegetables, or cosmic dust—we all dance to a mysterious tune, intoned in the distance by an invisible piper. Einstein

Contents

Introduction 1

Chapter 1 to Chapter 7 The Historical Era 3

1 The Discovery of Cosmic Radiation 3

1.1 Invisible Visitors from Outer Space – Cosmic Rays 3
1.2 The Mystery Begins – Why Does Charge Disappear? 5
1.3 Measuring Charge – The Electroscope 6
1.4 Corpuscular Particles Discovered - Electrons and Ions 8
1.5 The Discovery of Radioactivity and Its Effects 9
1.6 What Is The Origin Of The Mysterious Radiation? 11
1.7 Hess Balloon Flights 14

2 Confirmation of the Existence of Cosmic Rays 21

2.1 Theories, Explanations and Skeptics 21
2.2 Millikan's Balloon Experiments 24
2.3 Millikan's Lake Experiments 27
2.4 Final Confirmation 29

3 Atomic Structure and Cosmic Rays 31

3.1 Studying Cosmic Rays – A Complicated Problem 31
3.2 Atomic Structure 32
3.3 Radiation and Ionization 34
3.4 Interaction and Absorption of Radiation in Matter 35
3.5 Charged Particle Interactions with Matter 39
3.6 Photonic Interactions in Matter 43
3.7 Are Cosmic Rays Gamma Rays? 45

4 Early Theories, Developments and Conflicts 47

4.1 The Gamma Ray Assumption 47
4.2 Millikan's Theory of Cosmic Ray Formation 48
4.3 Cloud Chamber Sees a Cosmic Ray 49
4.4 Geiger Counters Transform Cosmic Ray Studies. 51
4.5 The Legendary Coincidence Experiments 55
4.6 Conflicting Hypotheses about the Primary Radiation 58

5 Using Earth's Magnetic Field to Study Cosmic Rays 63

5.1 Why Is the Earth's Magnetic Field Relevant? 63
5.2 Charged particles and magnetic fields 65
5.3 Searching For Global Cosmic Ray Variations. 68
5.4 Propagation of Charged Particles through Earth's Magnetic Field. 69

5.5 Discovery of the Latitude Variation. 74
5.6 Confirmation of the East-West Effect. 80
5.7 Role of the Atmosphere in Studying These Effects 82

6 Cosmic Rays and the Emergence of Elementary Particles 87

6.1 The Birth of Elementary Particle Physics 87
6.2 Discovery of the Positron and the Creation of Matter 92
6.3 Particle Showers and Interactions 98
6.4 Behavior of High Energy Electrons and Photons 102
6.5 Electromagnetic Cascade Showers 105
6.6 Discovery of Muons in Cosmic Rays 108
6.7 Properties of the Muon 112

7 Cosmic Rays Before and After World War II 121

7.1 Research Environment for Cosmic Ray Studies after the War 121
7.2 Doubts about the Current Muon Theory 124
7.2 Photographic Emulsions Come Of Age 129
7.3 Discovery of the Pion 134
7.4 Nuclear Active Particles 140
7.5 Radiation from Space 145
7.6 Completing the Cosmic Ray Picture 149

Chapter 8 to Chapter 15 The Modern Era 155

8 Cosmic Rays - From the 1950's to Today 155

8.1 The Cosmic Ray Field 155
8.2 Cosmic Rays Hitting Earth 159
8.3 Energies of Cosmic Rays 160
8.4 Detecting Cosmic Rays 162
8.5 Current Picture of Cosmic Rays 174

9 Origin of Cosmic Rays 177

9.1 Introduction 177
9.2 First Theories of Cosmic Ray Origins 178
9.3 Cosmic Rays from the Sun and Stars 180
9.4 Fermi's Acceleration Theory for Cosmic Rays 182
9.5 Supernova Theory of Cosmic Rays 184
9.6 Modern View of Cosmic Ray Origins 187
9.7 Ultra High Energy Gamma Rays and Cosmic Ray Origins 189

10 Large Cosmic Ray Experiments on Earth 193

10.1 Introduction 193
10.2 Akeno Giant Air Shower Array (AGASA) 193

10.3 Chicago Air Shower Array (CASA)....194
10.4 GAMMA Experiment....195
10.5 GRAPES-3 Experiment....196
10.6 High Altitude Water Cherenkov Experiment (HAWC)....198
10.7 High Energy Stereoscopic System (H.E.S.S.)....198
10.8 High Resolution Fly's Eye Project....199
10.9 Telescope Array Project....200
10.10 Pierre Auger Observatory....201
10.11 IceCube Neutrino Observatory....201
10.12 TAIGA Advanced Instrument Experiment....202
10.13 Large High Altitude Air Shower Observatory (LHAASO)....203
10.14 Spaceship Earth....204

11 Educational Cosmic Ray Projects....207

11.1 Introduction....207
11.2 Cosmic-Ray Extremely Distributed Observatory (CREDO)....207
11.3 HiSparc....208
11.4 Cosmic@Web....209
11.5 Extreme Energy Events (EEE)....210
11.6 QuarkNet....211
11.7 Washington Area Large-scale Time-coincidence Array....212
11.8 California High School Cosmic Ray Observatory (CHICOS)....212
11.9 Showers of Knowledge....213

12 Importance of Cosmic Rays to Our Planet....215

12.1 Introduction....215
12.2 Carbon 14....215
12.3 Biological Effects....218
12.3 Weather and Climate....219
12.4 How Cosmic Rays May Have Shaped Life.....220

13 Applications of Cosmic Rays....223

13.1 Introduction....223
13.2 Muon Detector Systems....225
13.3 Imaging the Core of Fukushima Reactor with Muons....226
13.4 Cargo Scanning with Muons....230
13.5 Decision Sciences Freeport Muon Inspection Station....231
13.6 ScanPyramids Project....233
13.7 Commercial Muon Systems....234
13.8 Other Projects Involving Muons....235
13.9 Measuring Soil Moisture with Cosmic Ray Neutrons....237

14 Cosmic Neutrinos....239

14.1 Introduction 239
14.2 Early History of Neutrinos 240
14.3 First Neutrino Detection 240
14.4 Solar Neutrinos 242
14.5 Neutrino Astronomy 244
14.6 IceCube Neutrino Observatory 245
14.7 IceCube Neutrino Observatory Hits the Jackpot 247

15 Too Interesting To Be Left Out 253

15.1 Introduction 253
15.2 NOvA as a Cosmic Ray Detector 253
15.3 Neutron Cosmic Ray Monitors 254
15.4 Hubble Telescope as a Cosmic Ray Detector 255
15.5 Using the Moon as a Cosmic Ray Detector 257

Appendix 1. Distances in the Universe 259

Appendix 2. Powers of Ten Notation 259

Appendix 3. The Electron Volt 260

Appendix 4. Energy, Force and Momentum 261

Appendix 5. Mass Per Unit Area 263

Appendix 6. Particle Physics 263

Appendix 7. Inverse Square Law. 264

Notable References 265

Glossary 267

Index 269

Introduction

In our daily lives we seldom ponder the wonderful and often mysterious nature of the universe. Look at the beautiful night sky. This starlight reaches us from the depths of the cosmos. However we are unaware of another cosmic energy hitting our planet just as continuously. This cosmic flux has the same intensity as starlight but cannot be detected with our ordinary senses. Yet it has profound consequences to life on Earth and our understanding of the physical universe.

During much of our history this cosmic rain, now known as *cosmic rays*, went undiscovered. But, little by little, scientists have pieced together a picture of this unusual phenomenon. While many of the details of cosmic rays have been discovered, some mysteries remain, and the picture is not yet complete.

For most of us, the story of the discovery of cosmic rays and the scientists that participated in the unraveling of the mystery is not well known. This is unfortunate, as the tale is full of adventure, intrigue, and surprises. Therefore, in the first seven chapters of this book, I have striven to recount the progress of the early scientists and their endeavors related to discovering cosmic rays. Some basic knowledge of science will be helpful here. To aid understanding, a section on atomic structure has been included. Wherever possible, complex concepts are handled without equations. I have tried to make it interesting for readers through the use of numerous photos and illustrations. Appendices are provided for those who need refreshing on some technical details.

Cosmic rays are still being studied rigorously today. Networks of monitoring stations are in operation with data being recorded on a daily basis around the world. There are also numerous educational programs and monitoring sites at schools and other institutions in many cities. Hopefully these programs will encourage students to learn more about this subject.

Research methods using new and more sophisticated methods are being employed to study the most energetic cosmic rays. We will take a look at several of these that are in operation around the globe.

What we currently know about the properties of cosmic rays and methods for studying them are important for future studies. At this point a great deal is known about their energy and some other properties. But, many effects of cosmic rays on our planet are still being discovered. Later on in the book, we'll take a look at some of these connections.

Science often moves ahead with spurts of experimental work and theoretical advances. The field of cosmic rays is no different. In recent years many ideas, only imagined in this field, have become an operable possibility. For example the dreams of applying cosmic rays to image the interior of active volcanoes, nuclear reactors, or detecting nuclear contraband have been realized. Technical advances in many fields have made this possible.

Scientists are also looking at the emerging field of cosmic neutrino astronomy. These and other ideas may shed light how cosmic rays fit into our understanding of dark matter, dark energy and the birth of the universe. It will be interesting to see how the entire picture emerges.

The author is indebted to the many experts that have paved the way for the material in this book. There is an enormous wealth of knowledge in the books and papers of these eminent scientists.

And finally I need to mention the look of awe on the faces of students when they see my cosmic ray telescope displaying a cosmic bullet that has come right through the roof of the building!

George R. Steber

Chapter 1 to Chapter 7 The Historical Era

1 The Discovery of Cosmic Radiation

1.1 Invisible Visitors from Outer Space – Cosmic Rays

Hitting Earth every second from all directions is a bombardment of invisible cosmic bullets that penetrate our atmosphere – having many effects on our planet (Figure 1.1). Cosmic rays – as we now call them – contain the most powerful, penetrating particles known to mankind, possessing far greater energy than any that can be generated by the most powerful atom smashing machines that scientists on Earth can build.

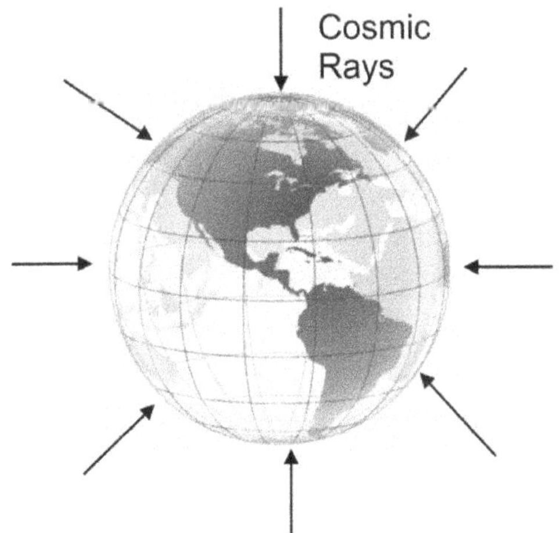

Figure 1.1 Earth is constantly bombarded by cosmic bullets from outer space.

To some extent our atmosphere and geomagnetic field shields us from these particles. The atmosphere absorbs some of them and the magnetic field diverts the weaker primary rays but most of the higher energy ones collide with air molecules in the stratosphere and generate secondary rays causing a cosmic rain in the atmosphere, (Figure 1.2). Cosmic rain, unlike ordinary rain, is very powerful and it can travel to the Earth's surface and even penetrate the surface to the depth of coalmines or the bottom of lakes. These particles also pass through us giving us a small but unavoidable dose of radiation – not considered harmful at sea level.

Just what are these mysterious visitors from outer space and where do they come from? We have studied them for well over one hundred years and it is still not known for certain where they originate or how they were created. Indeed there are a multitude of questions about their effects on our planet and biosphere. Yes – these invisible cosmic visitors are in many ways profoundly mystifying.

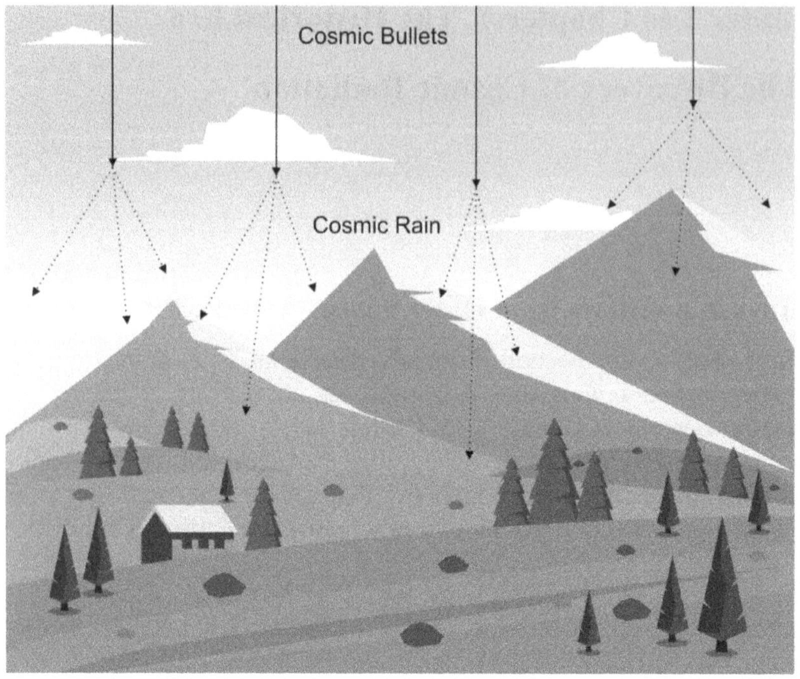

Figure 1.2 Cosmic bullets hitting the upper atmosphere create cosmic rain on the surface of the Earth - penetrating to the depth of coalmines or the bottom of lakes.

Over the last hundred years, considerable information has been deduced about the *physical nature* of cosmic rays. Much is now known about their energy distribution, particle composition and possible sources. Studies of cosmic rays led to the discovery of the first anti-matter particle as well as a host of other particles such as pions and muons, which we will learn more about later on. More profoundly, cosmic ray studies have helped us understand the physical nature of the Universe.

Our understanding of cosmic rays may be related to mysterious objects in our Universe like supernovae, blazars and black holes, together with the enigmatic dark energy and dark matter.

Here on Earth, cosmic rays influence our daily lives in many ways that we do not fully understand. They act upon our atmosphere and affect our health. New uses for cosmic rays are constantly being discovered such as imaging of volcanoes and searching for nuclear contraband. We will learn a lot more about these things later on in the book. There is still a long way to go and much to be discovered. It promises to be an interesting journey.

How did we first learn about cosmic rays? The circumstances leading to their discovery is not well known. That is unfortunate since it is an interesting story and illustrates how discoveries in one area of science can profoundly affect other areas. In what follows we will try to take you through the major historical events related to the discovery of cosmic rays. The story is full of surprises, false claims and erroneous theories with scientists from all over the planet participating in the mysterious adventure.

1.2 The Mystery Begins – Why Does Charge Disappear?

Surprisingly, the discovery of cosmic rays owes its beginnings to a completely different field – the study of electrostatics. Static electricity has been known for centuries. It is the voltage you generate when you walk across a carpet on a dry day. *Triboelectric* generation, as it's called, can create very high voltages, ranging to thousands of volts. As you are undoubtedly aware, discharging the electricity by touching a metal object will often give you a nasty shock. The voltage is generated because the atoms on the bottom of your shoes are rubbing against the carpet and removing some electrons.

When tribocharging happens, an unbalanced situation is created; electrons are detached from their atoms and the carpet becomes positively charged. This separation of charges is responsible for the ensuing voltage. In a discharge, the recombination of the electrons with their positive counter-parts results in the electric spark.

Figure 1.3 A Van de Graaff voltage generator will make your hair stand on end!

Touching a high voltage Van de Graaff generator, while standing on an insulator, will also transfer charges to your body (Figure 1.3). Since the charges will distribute themselves evenly throughout your body, this can make your hair separate and stand on end. It is a good example of *like* charges repelling.

1.3 Measuring Charge – The Electroscope

A crucial invention for the study of charged objects is the electroscope. It plays a major role in the discovery of cosmic rays. The Frenchman Jean Nollet invented the electroscope in 1746. The Englishman, Abraham Bennet in 1787, invented the more familiar gold leaf electroscope available in some high school laboratories today, (Figure 1.4).

In its simplest form, a gold leaf electroscope consists of two extremely thin gold leaves hanging from a metal rod inside of a metal case,. The metal charging rod protrudes through the case through a hole in an *insulator*. An insulator is a material that does not conduct electricity and is used to isolate the rod from the case. The metal case, which is normally grounded, is used to protect the leaves from air movement and extraneous electric charges. The gold leaves are usually viewed through a glass window. When a charged object is touched to the top of the sensing rod some of the charge is transferred to the rod where it spreads itself all over the rod and gold leaves. This causes the leaves to fly apart to form an inverted V. The leaves fly apart since the same quantity of charge is on both gold leaves and are thus subject to a repelling force. This works for both positive and negative charges, as electroscopes are not capable of telling the difference. Let's take a closer look at this phenomenon.

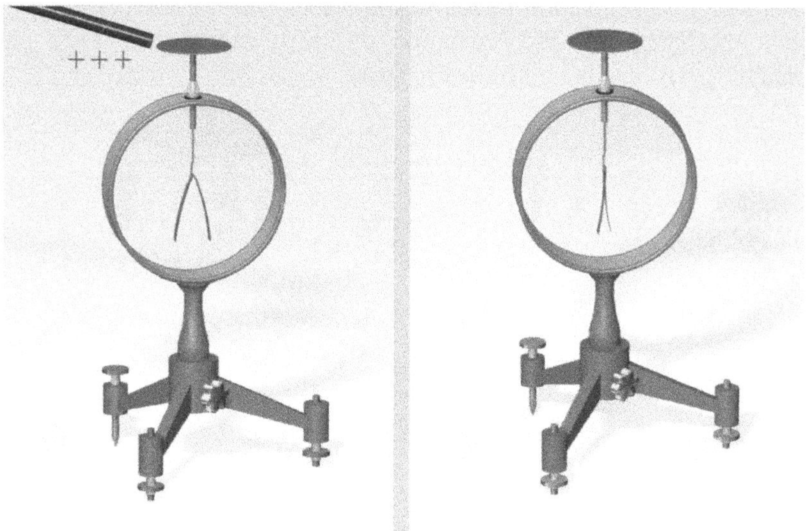

Figure 1.4 Gold leaf electroscope. Left is charged. Right is discharged.

The physical process of leakage is observed when the electroscope is charged and allowed to remain in that state for a period of time. Initially the leaves are apart since each leaf has the same charge. But very slowly they drop to their original position as the charge dribbles away. Normally, in electroscopes, it is assumed that the leakage charge escapes through the imperfect insulator holding the rod. Using the right materials and airtight construction should be able to reduce this leakage to near zero. In spite of these precautions leakage was still observed in electroscopes. For more than a century scientists were aware that even the best electroscopes could not hold a charge indefinitely. But no creditable theory to explain it could be found.

Around the eighteenth century, people began experimenting seriously with static electricity. Most notable was a Frenchman named Charles Coulomb who derived the basic law that applies to electrical charges and forces between them. His famous experiment showed that electric repulsion between *like* charges obeys a law having the same form as Newton's law of gravity – an *inverse* square law. (An inverse square law is a purely geometrical effect. It shows that the intensity of a point source decreases with the inverse square of distance.)

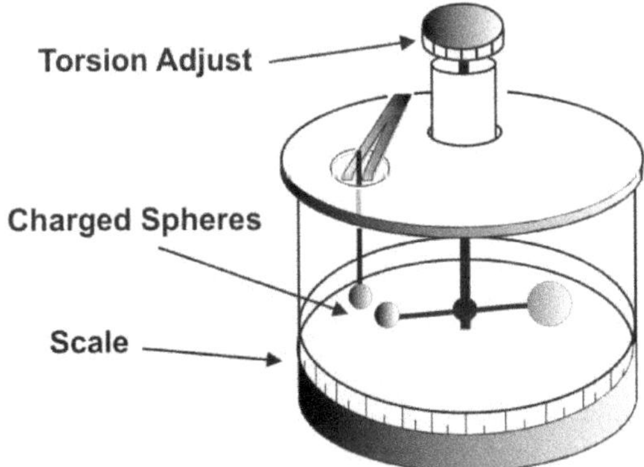

Figure 1.5 Coulomb's charge measuring apparatus using two charged spheres.

Coulomb's experimental device used two charged metal spheres, one fixed and one on the needle arm of a torsion balance, (Figure 1.5). Extraordinarily small forces between the balls could be measured with this apparatus. It works this way. After bringing the two balls together at the zero degree mark of the scale, they are grounded. Then – while they are still in contact – the two balls are electrified. By symmetry, the charge distributes itself equally onto both balls and they repulse each other. Following a few oscillations of the ball on the needle arm, it settles at a distance away from the stationary ball and its distance is recorded. Next the micrometer is twisted an amount, which quadruples the total force of torsion. The distance is examined to see if the distance between the balls converges to half, as would be the case if it were an inverse square law.

In this procedure much time was devoted to charging these metal spheres. Coulomb noticed that over time the spheres would lose their charge and not be repelled. Many things were tried in an attempt to reduce the charge loss but it could not be eliminated completely. Most importantly, it was observed that the leakage depended on the influence of air. For example the discharge rate was higher on days of high humidity. On dry days the charge would remain much longer. But in this case it too would eventually discharge. It was assumed that the construction details of the apparatus and air conditions caused *all* the leakage. Many years later it was found that there was something mysterious in the air causing the discharge.

1.4 Corpuscular Particles Discovered - Electrons and Ions

Near the late nineteenth century a topic of scientific investigation was the conductivity of gases and their ability to conduct electricity. A milestone occurred in 1897 when an Englishman J. J. Thompson (a.k.a. Thomson) produced cathode rays in an electrified gas tube. These rays were *corpuscular* in nature and behaved like small particles. Indeed he had found a new particle – of very little mass – having a negative electric charge. Thompson called the particles "corpuscles", but later scientists preferred the name *electron*. The discovery of the electron was a major event in physics.

At this point scientists were learning more about the structure of matter. They knew that matter consists of atoms and that different types of atoms are the basis of chemical elements. These elements when combined properly can, in turn, form molecules of various substances. They knew too that atoms have an equal balance of positive and negative charges and are essentially neutral. And, as Thompson showed, charges themselves show a granular structure and may be set free from the atom under the right conditions.

It was thought that the discharge of the electroscope could be explained by assuming that gas molecules somehow lose an electron and become positively charged. This process is called *ionization*. The ion pair thus formed consists of the negative electron and the positive molecule, or ion, as it is known. According to this picture, the negative electron floats around free and eventually attaches itself to a positively charged ion. Experiments had shown that the air inside the electroscope was always slightly ionized with small amounts of electrons and charged ions. In theory the positive charged gold leaves of the electroscope will attract negative ions and repel positive ions. Consequently the positive charge on the leaves will gradually be neutralized and the leaves will drop. If we initially use a negative charge on the leaves the same will be true. In this case the positive ions are drawn to the leaves and are neutralized. However no explanation was forthcoming for the slight ionization that was always present in the air of the electroscope - until *radiation* was discovered.

1.5 The Discovery of Radioactivity and Its Effects

Scientists discovered various types of radiation in the late 1800's. This was a golden age of discovery. In 1895 Wilhelm Roentgen discovered x-rays and found that they could discharge an electroscope. Then in 1896 Henri Becquerel discovered natural radioactivity while studying the photographic images of various fluorescent and phosphorescent materials. Pierre and Marie Curie discovered radioactive polonium and radium in 1898. Then in 1899 Ernest Rutherford identified two distinct kinds of rays emitted by uranium. Common to all of these discoveries is that scientists now knew that these types of radiation were powerful enough to ionize gases and *discharge* an electroscope.

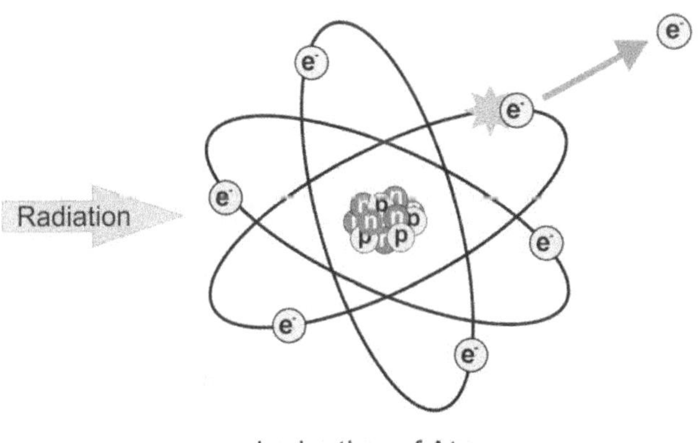

Figure 1.6 Ionizing radiation entering on the left energizes the atom causing it to lose an electron as shown on right. Atom becomes a positive ion.

Ionization happens because the radiation emissions strip electrons off of the gas molecules making them positively charged ions (Figure 1.6). This trail of ions provides the discharge path for the electroscope. Many more ions are formed for a stronger radiation source than a weaker one. Hence the electroscope will discharge faster for strong sources than weaker ones. This observation provided an interesting application for the electroscope. It could now measure the intensity of radiation by looking at the charge leakage rate. In this way different radioactive materials could be classified as to strength or intensity of emission.

Everything seemed to be understandable in this procedure except for the same puzzling fact. When obvious sources of ionization were removed, the electric charge continued to leak away in the electroscope. Now most electroscopes at that time operated with air as the working gas. So the problem of charge leakage was explained as being due to charged dust in the air, which formed a leakage path. If this were true, what caused the dust to be charged?

Figure 1.7 Charles Wilson was one of the first to measure gold leaf electroscope leakage rates. Later on he became famous for the invention of the Cloud Chamber.

Around 1900, a Scottish scientist, Charles Wilson (a.k.a., C. T. R. Wilson) (Figure 1.7) of Sydney Sussex College in Cambridge, England, decided to accurately measure the leakage rate of a gold leaf electroscope. He used very well insulated parts and dust free air. As we have noted above, when charged, the gold leaves in the electroscope are spread apart, then slowly come back together as the charge leaks away. He measured the rate of charge loss by precisely observing the gold leaves. Amazingly, they *always* showed a residual leakage that was the same in daylight and in darkness, and for positive or negative charge.

In describing his results he made the profound statement: "*Experiments were now carried out to test whether the production of ions in dust-free air could be explained as being due to radiation from sources... outside our atmosphere, possibly radiation... but of enormously greater penetrating power.*"

1.6 What Is The Origin Of The Mysterious Radiation?

Wilson's discovery of residual ionization puzzled him. If the radiation causing it was of cosmic origin then moving his apparatus higher in the atmosphere or lower in the ground should produce a marked change in the discharge rate. At that time, 1901, it was difficult to go higher in the atmosphere, so he chose to go underground. He set up his apparatus in Peebles in Scotland in a Caledonian Railway tunnel. Unfortunately for Wilson, the Earth is also a source of radiation and his electroscope was unable to distinguish between cosmic and Earth radiation. After some disappointing results he abandoned this work and assumed, albeit incorrectly, that the residual ionization was due to some process inherent in the air.

Most scientists at the time were in agreement with Wilson and considered the observed residual ionization as due to weak radiation. Some guessed that it might be caused by small traces of radioactive material in the electroscope itself. But even when this was taken into account, a large amount of residual ionization remained.

Around 1903 two Canadian groups, McLennan and Burton from the University of Toronto and Rutherford and Cooke from McGill University noticed that the rate of discharge of an electroscope could be reduced by as much as 30% by enclosing it in an air tight, thick metal shield. This meant that the leakage of the electroscope was not entirely due to poor construction of the insulator. Instead it must be due to some highly *penetrating rays* like gamma rays given off by radium.

It should be mentioned at this point that the production of ions in a volume of gas inside an electroscope chamber is measured in *ions per cubic centimeter per second*. For normal air at sea level this number lies between 10 and 20. If the chamber is protected by lead this number will be reduced. In order to evaluate results, physicists of the day used ions/cc/sec as a reference for comparison.

For several years after Wilson's experiment, the consensus of scientists attributed the weak radiation to radioactive material in the earth's soil. This assumption is in agreement with Wilson's railroad tunnel result. At this point many scientists realized that comparing electroscope observations at different altitudes could easily test this premise. If the radiation came from the soil it should be more intense near the surface and gradually diminish with height. To test this idea electroscopes were carried to the tops of tall buildings but, due to the poor electroscopes of the day, the results were inconclusive and attributed to radioactive material in the buildings.

Father Theodor Wulf (a.k.a. Thomas Wulf) (Figure 1.8) of Holland and Rome performed one of the more interesting experiments. In 1909 he had perfected a new type of electroscope, replacing the gold leaves by two very thin wires held under tension by a thin quartz fiber. His more sensitive and more robust portable electroscope was then manufactured by the German instrument firm Günther and Tegetmeyer.

Figure 1.8 Theodor Wulf, a German Jesuit, studied physics in Innsbruck and Göttingen. He improved the electrometer's design by replacing the thin metal foils with two thin metal fibres. This new electrometer was much easier to calibrate and became the state-of-the-art detector for outdoor measurements.

As shown in the Figure 1.9, Wulf's instrument consisted of a 17-cm-diameter zinc cylinder with a pair of flexible wires below the access tower. The wires are pushed apart by static electricity, and the microscope (looking in from the right) measures their separation, illuminated by light from the mirror at left. The air in the cylinder was kept dry by sodium in the small recess below the microscope.

In 1910, Wulf brought some of his new electroscopes with him on an Easter holiday visit to Paris. Hoping to end the controversy, he ascended 900 feet up the Eiffel tower with his electroscopes. But, alas, he was disappointed. While he did observe a decrease in the leakage rate of 64 percent it wasn't as low as was expected.

If the radiation was coming *solely* from the ground much more of it was expected to be absorbed by the air between the surface and the top of tower. Realizing his electroscope measured the combined radiation from the Earth and atmosphere, Wulf wondered if the Earth's radiation was actually decreasing as fast as predicted but somehow it was being overshadowed by another source of radiation coming from the atmosphere. However, without more proof his assertions were not given general acceptance.

Between 1909 and 1910 a Swiss physicist, Albert Gockel began making balloon flights to study the variation of radiation with altitude using an electroscope of the Wulf type. This was the first significant airborne experiment at this point. His first flight occurred on December 11, 1909 and carried with him a meteorologist Dr. Quervain and pilot Lieutenant Muller. It was a four-hour balloon flight that ascended to a height of 13,000 feet. In a modest attempt to protect his electroscope, Gockel sealed it with leather and grease but regrettably not from the atmosphere.

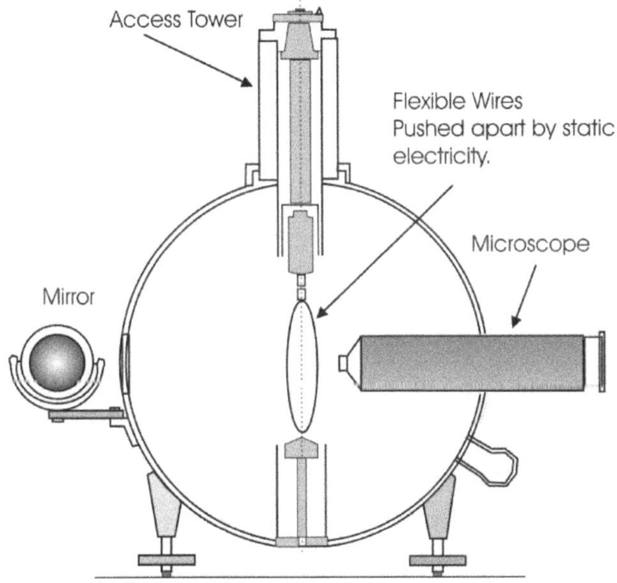

Figure 1.9 Wulf's Improved Electroscope was more complicated and accurate than other models and used flexible wires and a microscope.

He reported that the *penetrating radiation* was equally as large at this altitude as at the Earth's surface. This was again in conflict with current radiation theory that predicted that it should fall to one-half its surface value as early as 250 feet. He concluded, as had Wulf, that there must be another source of radiation. It should be noted that a major mistake made by Gockel was in failing to compensate for pressure variations with altitude in the electroscope. Had he done so, he may have seen a radiation *increase* with altitude.

It is interesting that both Wulf and Gockel showed that the radiation did not *decrease* with height as expected with the theory of the time. As happens sometimes in science, both Gockel and Wulf had stumbled on the correct result but could not make sense of it; nor could other scientists who in 1909, 1910 and 1911 made balloon ascents to measure ionization levels.

1.7 Hess Balloon Flights

In 1911 an Austrian physicist, Victor Hess, became interested in the problem of variation of radiation with altitude and began the work that 25 years later would make him a Nobel Prize recipient.

Figure 1.10 Hess's balloon, named the *Böhmen* (German for Bohemia). It towered twelve stories high with a lifting power of two tons.

While reading about these earlier experiments, Hess speculated as to whether the source of ionization could be located in the sky rather than the ground. That he was an ardent balloonist (Figure 1.10) was a happy circumstance. In preparation for his balloon ascents, he estimated the height at which ground radiation would stop producing ionization, about 1500 feet, according to current rules of radiation attenuation. (Since radiation interacts with matter, its intensity will decrease as it travels through a material. At the time, x-rays and gammas were known to diminish with distance according to an exponential law.)

Hess made ten balloon ascents, including five at night. With him in the basket were a meteorologist and a pilot (Figure 1.11). The first two flights in 1911 lasted 4 hours and went up to a height of 3,500 feet. During both of these flights, the radiation was measured to be constant with altitude. This was unexpected because as we have seen before, the radiation was thought to diminish with height. Were his instruments faulty?

Figure 1.11 Victor Hess shown in the gondola of his hydrogen filled balloon.

The next year, in August 1912 there were seven balloon flights. Special care was taken to design instruments that would not be damaged by temperature and pressure changes. His electroscopes were of the Wulf type but very precisely constructed (Figure 1.12). Two of the three electroscopes had airtight cases so the internal pressure would remain constant. This allowed his instruments to be independent of altitude. Unlike Gockel he realized that the sensitivity of unpressurized electroscopes decreases with altitude due to the lower ionization ability in a given volume of air.

During the first six flights the balloon did not reach sufficient altitude. Before the famous seventh flight on August 7, 1912 near Aussig in Austria, they filled the balloon "Böhmen" with hydrogen instead of coal gas and were propelled over the next few hours to over 16,000 feet. Captain W. Hoffery was the pilot and the meteorological observer was W. Wolf. Hess listed himself as “observer for atmospheric electricity". It was a daring flight - riding high in the sky in a small basket tied to a highly flammable container of hydrogen gas. Their flight was all the more adventurous since it was without an oxygen supply for breathing in the thin air.

Ascending through the atmosphere Hess found the ionization slightly weaker at first as indicated by the slower discharge rate of the electroscopes. This is in accordance with the expected radiation coming from the Earth's surface. It diminishes with distance. Once he arrived above 2000 feet the trend reversed and ionization began to increase. It continued this trend with altitude as though the balloon was approaching a definite source of radiation. At 16,000 feet (4.87 km) the electroscopes were discharging at 4 times the rate on the surface (Figure 1.13).

Figure 1.12 One of the electrometers used by Hess during his famous flight.

Later interpretation by Hess concluded that the radiation from Earth had an influence up to about 2000 feet (0.61 km) but, after that, at higher altitudes there was another source of radiation. He then made a bold assertion: "*A radiation of very high penetrating power enters our atmosphere from above.*"

In attempting to determine the source of the radiation, Hess considered the sun. He took balloon flights at night and did not notice any reduction in ionization compared to daytime flights. Once he made an ascent during an almost total eclipse of the sun on April 12, 1912. The ionization did not decrease during the eclipse. He concluded that the sun could not itself be the main source of the radiation. Today we know that cosmic rays arrive at the Earth isotropically from space with the sun contributing its part. In any case, the source of the penetrating radiation remained a mystery.

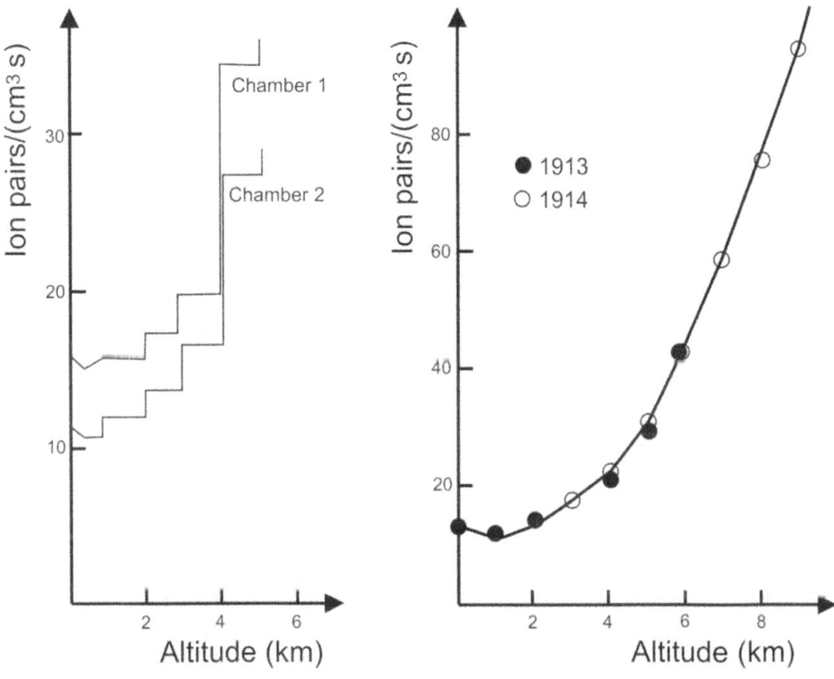

Figure 1.13 Increase of ionization with altitude as measured by Hess in 1912 (left) and by Kolhörster (right) in 1913. Note different vertical scales.

Another scientist, W. Kolhörster made brave balloon flights in 1913 and 1914 reaching altitudes of 28,000 feet (Figure 1.14). His Wulf type electrometers were much better than those of Hess and were tested for the effects of temperature. At this altitude he found ionization levels even stronger than found by Hess.

Hess's discovery and Kolhörster's confirmation did not find acceptance among fellow scientists, at least not initially. In fact there was vehement opposition, controversy and heated discourse on the subject.

Scientists, you know, have strong convictions. Things got so bad that for a time it was not even fitting to speak about the subject of this mysterious radiation, as it seemed more like witchcraft than physics. Many years would pass before it would be generally accepted.

Figure 1.14 Werner Kolhörster in his gondola standing behind the pilot.

As sometimes happens in the world of science, there is an uncredited pioneer in this story. Domenico Pacini (Figure 1.15), then a researcher in Rome was also studying the subject of penetrating radiation. Pacini proposed and carried out novel experiments to observe radiation strength under water between 1907 and 1911. Some experiments were on the sea and others on lakes.

He reported that the radiation at ten feet (3 m) underwater was much less (20%) then the surface radiation. This showed conclusively that the penetrating radiation was decreasing with depth and could not come from the Earth. He published his findings (in Italian) in a note titled: "Penetrating Radiation at the Surface of and in Water". Therein he reported: "*Observations carried out on the sea during the year 1910 led me to conclude that a significant portion of pervasive radiation that is found in the air... was independent ... of the active substances... of the Earth's surface.*" This was one year before the famous balloon flight.

Figure 1.15 Domenico Pacini is seen observing with an electroscope, similar to the one used to monitor radiation at 3 meters deep at Lake Bracciano Italy.

Some years later in 1920, Pacini wrote to Hess mentioning his sea results and wondering why no reference of them was in Hess's papers. After the exchange of several letters Hess acknowledged he was familiar with the sea measurements and that Pacini certainly had priority for the measurements at sea. However, Hess maintained, in his reply, that "*the demonstration of the existence of a new source of penetrating radiation from above*" was his discovery.

With the turmoil of World War I approaching in 1914 a great decline in scientific activity occurred and cosmic ray research fell into stagnancy. This inactivity lasted between 1915 and 1925 with no significant contributions to the subject during this period.

Chapter 2

2 Confirmation of the Existence of Cosmic Rays

2.1 Theories, Explanations and Skeptics

After Hess's controversial finding in 1912 that cosmic radiation increased with altitude, little was done to identify the type of radiation. The scientists of that day can be excused for this, as there was little in the way of scientific equipment available for any studies. And unlike terrestrial radiation, cosmic rays could not be controlled for experimental purposes. Also, the amount of cosmic radiation, measured at the surface, was small compared to terrestrial radiation.

"Professor Hess, no one seems to know what your balloon flight means."

But even beyond these difficulties most scientists of the day were not eager to study cosmic rays further; they thought they already knew all there was to know. So the field was in stagnancy until Millikan's work 16 years later.

In science the results of one experiment are usually not enough to establish a new theory. There could be flaws in the measurements or the procedure could be faulty. Hess's results and bold assertions needed to be verified. There were numerous questions yet to be answered. Understandably many scientists were reluctant to accept Hess's result and especially the interpretation of his experiments without further proof.

First of all they needed to confirm the results. Subsequent balloon flights by Werner Kolhörster of Berlin were especially significant. His daring flights between 1913 and 1914 ascended above 16,000 feet to a maximum of 28,000 feet. There he found ionizing radiation far stronger than that reported by Hess. In fact it was fifty times the value on the surface. This was a significant confirmation that radiation varied with altitude.

But there were still questions about the uncertainties and accuracy of the instruments and indeed many physicists remained skeptical. To accept a theory, it is incumbent on us to explore all other possible explanations for the phenomenon, in this case the unexplained penetrating radiation source.

Another scientist at the time, King, had serious reservations. He claimed that Hess's high radiation readings at high altitude were due to an entirely different phenomenon. He believed that the balloon was acquiring many thousands of volts in its upward journey through the accumulation of static electricity, much like we would obtain by walking across a carpet on a dry day. This voltage would surely act as a great attractor to ions in the atmosphere. These ions supposedly would cause the electroscope to discharge faster.

But this theory is not correct. We know that ions are just charged molecules of air and as such could not penetrate the sealed walls of the electroscope. The ionization measured is created *in situ* by something else penetrating the walls. Therefore the electroscope would continue its function faithfully regardless of how many ions are drawn to the balloon.

Charles Wilson, who we met earlier, also had a different theory. Wilson, who by the way, became famous for the invention of the cloud chamber, strongly disagreed with Hess's interpretation suggesting that he was overlooking an obvious source of the radiation. Based on his experience as an expert in ionization processes, Wilson suggested that the radiation was produced by thunderstorms at the top of the atmosphere. He supposed that electrons in thunderstorm clouds could be accelerated to very high energies and produce radiation. This would explain why the radiation increased with altitude.

Other scientists thought that the atmosphere might contain small amounts of radioactive gases. At the time, some radioactive gases like radon were known to exist as a gas. Obviously if these gases rose to the higher layers of the atmosphere, the radiation would be explained.

Upon closer examination, both of these explanations had flaws. Both implied that the radiation effect would vary with the weather conditions. And since these conditions are constantly changing, the radiation should vary with the time of day and the season. This clearly is the case with thunderstorms since they do not happen all the time and the hypothetical radiation they would produce would not be constant. Similarly, the hypothetical radioactive atmospheric gas would not stay constant through changing weather conditions.

Hess's measurements had indicated a *constant* flux of radiation. It was steady by day or night, in rain or shine, and did not vary with the seasons. At least that's what the experiment showed within the limits of its accuracy. Hence it was barely imaginable that these explanations were correct. We know *now* that there *are* slight variations in cosmic radiation due to a variety of effects.

There was an interesting turn of events here. The Swiss scientist, Albert Gockel (Figure 2.1) who had made the first balloon flight in 1910 to 4.5 km had also decided to study the radiation increase with altitude but from mountains far removed from potentially radioactive rocks by staying on the ice from deep glaciers. He published his work on radiation measurements on the following mountain peaks, 1909 Rothurn (2.3 km), 1911 Zermatt (2.6 km) and 1912 Matterhorn (3.3 km). Measurements made at these mountain peaks convinced Gockel that there should be radiation coming from space.

Figure 2.1 Albert Gockel was a professor (University of Fribourg) who in 1910 measured radiation at high altitudes. He discovered an unexpected dependency of radiation that it did not decrease with altitude, as one would expect if radiation is coming only from the Earth.

He confirmed the results of Kolhörster reporting a leakage rate greater at altitude than at the surface of the Earth. In the ensuing paper it was implied, very definitely, that the mysterious radiation was not caused by radioactivity but something above the atmosphere.

The publication of Gockel's paper stirred up a storm of criticism. A number of scientists of the day, Hess among them, went on record as opposed to this idea of cosmic origin. To be fair, cosmic origins at that time meant sources like the sun, moon, nearby stars and perhaps our own galaxy. Hess and others had found no evidence of diurnal variations due to the rotation of the Earth, as would be expected from heavenly bodies as they pass through our window of the sky. No one, at the time, ever dreamed of interstellar space as the origin.

2.2 Millikan's Balloon Experiments

Physicists are a skeptical group. By nature, they question results and search for answers. This situation was no different. Most skeptical at the time was Robert Millikan, then a prominent experimental physics professor at the California Institute of Technology (Figure 2.2). (Millikan is best known for measuring the charge on a single electron. In 1908, while a professor at the University of Chicago, Millikan worked on his famous oil-drop experiment in which he measured the charge on a single electron and derived the mass. As we saw earlier, J.J. Thomson had already discovered the electron. However, the actual charge and mass values were unknown.)

Figure 2.2 Physics professor Robert Millikan in his laboratory at Caltech.

Millikan, Ira Bowen and their team decided to investigate the claims of Hess. Foremost among the tasks was to obtain their own experimental data and to determine if there were any reasons to believe that radiation was arriving here from space. They decided to explore the radiation at very high altitudes through a series of extraordinary experiments.

In the spring of 1922 at Kelly Field, near San Antonio, Texas they sent up sounding balloons to over 50,000 feet or nearly ten miles. This is twice the height previously attained. Their instruments were marvels of technology. His exceptionally well built electroscopes were state of the art for ruggedness and sensitivity (Figure 2.3).

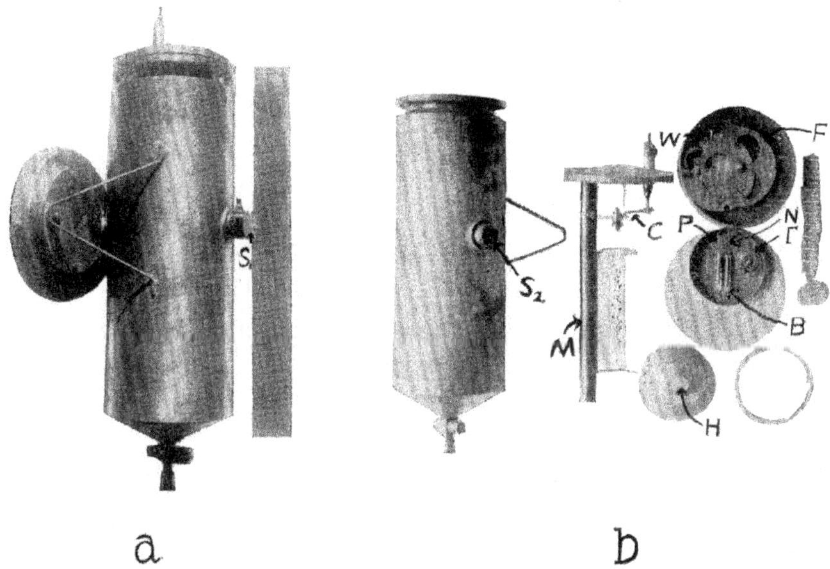

Figure 2.3 Shown in (a) is the complete recording instrument ready for flight. A 6 inch (15 cm) rule is shown next to it for comparison. Shown in (b) is the inner mechanism. The electroscopes were made of 0.3mm thick steel and designed to support a pressure of 10 atmospheres.They carried photographic film and the necessary driving mechanism for obtaining a continuous record of electroscope data, temperature and pressure during the ascent and descent. The electroscope consisted of two sputtered quartz fibres as used in the Wulf electroscope. Deflection was produced by a 300 volt battery. A shadow image (diffraction pattern) of the fibres produced by the light from the sky passed through an exceedingly fine vertical slit S1, shown at the left of the rule, fell upon the photographic film. The light tight slit H was over the thermometer and barometer chamber which registered on the circular film by sky lighting. The total weight with all contents was but 190 grams, or about 7 ounces.

The apparatus was built of steel to hold 300 cc. of air at 150 lbs of pressure and were provided with a recording barometer, thermometer and electroscope with two sets of moving photographic films. Together with the driving mechanism, the total weight was only 7 ounces. Also unusual for the time was that the altitude was not only established with well known sounding balloon laws but also by theodolite observations provided by the U. S. Signal Corps. His team chose to use unmanned flight because of the recognized danger of manned flight like that used by Hess and Kolhörster. Using unmanned flights eliminated the danger and the cost of human flight.

Perhaps most unique was the use of two balloons for lifting the instrument. In Millikan's words, "The purpose of sending up two balloons with each instrument is not merely to gain additional buoyancy but to enable one of them to bring the instrument safely to Earth after the other has burst and also to serve, after the return to Earth, to attract attention and increase the likelihood of the return of the instrument to the sender as per instructions attached to it."

During the flight, automatic recording devices would be employed to capture the results on photographic film. At the desired altitude, one of the balloons would burst causing the apparatus to descend slowly back to Earth. The apparatus was then recovered and the data processed.

Millikan said later that they expected – if the results previously reported were correct – to find very large discharge rates because their instruments went up through 88 percent of the atmosphere leaving only 12 percent left to absorb the intensity of the hypothetical rays entering from outside. Their electroscopes should have been exposed to radiation approaching the intensity at the top of the atmosphere.

At 5 km altitude Millikan found a mean discharge rate of the electroscopes at about 3 times that at the surface. Millikan said, "*This shows quite unambiguously, in agreement with the findings of Gockel, Hess, and Kolhörster, that the discharge rates at high altitudes are larger than those found at the surface.*" But Millikan also reported, "*We actually failed to find anything like the computed rates of discharge*" of those researchers in Europe.

He said further that his results proved conclusively that "*a radiation of the assumed properties did not exist, our observed rates of discharge being no more than one-fourth the computed amounts.*" In other words, Millikan was convinced that the radiation was local, because its intensity over Europe was four times that over San Antonio, Texas where his tests were performed. We now know that the Earth's magnetic field allows more cosmic rays into the atmosphere the closer you are to the magnetic poles. The magnetic field is quite different over Texas than it is over Europe where Hess made his measurements.

Since the origin of the penetrating rays was still uncertain Millikan and a colleague, Russell Otis, in the summer of 1923 went to the top of Pike's Peak. They carried 300 pounds of lead and a large 216 cubic foot tank of water to perform absorption experiments. Their Pike's Peak observations showed "*that if rays of cosmic origin existed at all they must be of different characteristics from any yet suggested...*" This experiment failed to decide if the radiation was from the heavens or some terrestrial cause, such as electrical discharges from thunderstorms or radioactivity.

2.3 Millikan's Lake Experiments

Continuing with the work, in the summer of 1925 Millikan along with Harvey Cameron (Figure 2.4) planned some new experiments to settle the question of the existence of a penetrating radiation of cosmic origin and to throw light on the cause of the variation of radiation with altitude.

Figure 2.4 Dr. Millikan (left) and Dr. G. Harvey Cameron with electroscopes they sank in California and Bolivia mountains lakes to detect cosmic radiation. The instruments were raised and examined through the eyepiece.

They proposed to use immense quantities of water as an absorbing material and chose two very deep snow-fed lakes to reduce the possibility of radioactive contamination. The first experiments were at Muir Lake (11,800 feet high) and close to Mount Whitney, the highest peak in the United States. They worked for ten days sinking their electroscopes to great depths. To their amazement, the electroscope readings kept decreasing down to 50 feet. The atmosphere above the lake is equivalent to 23 feet of water so the total absorption was equivalent to 73 feet of water, or equivalently 6 feet of lead. This penetration far surpassed any of the known radiation sources of the day.

To fully confirm that these rays were of cosmic origin Millikan's group went to the second lake, Arrowhead Lake, 300 miles farther south in the San Bernardino Mountains. Arrowhead Lake is at 5,100 feet and consequently 6,700 feet lower than Muir Lake. This height difference, 6,700 feet of air, represents an absorbing mass equal to 6 feet of water.

The experiment was repeated and when the data was analyzed it showed that every reading at Arrowhead Lake corresponded to a reading 6 feet farther down in Muir Lake. This is shown in the data plotted in the accompanying Figure 2.5. Based on these measurements, Millikan now changed his opinion and stated that this shows: "*that the rays do come in definitely from above, and that their origin is entirely outside the layer of the atmosphere between the two lakes.*"

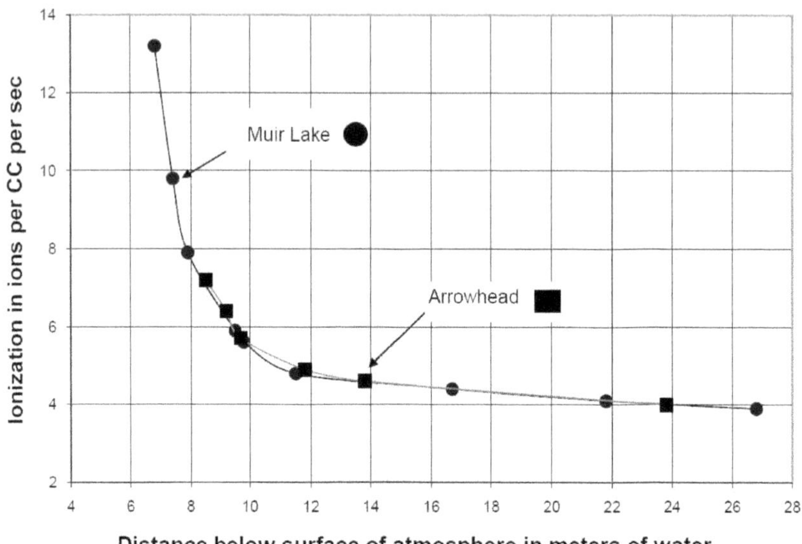

Figure 2.5 Millikan's ionization readings at Muir Lake and Arrowhead.

As a technical note it should be mentioned that Millikan's argument was not entirely correct. He assumed that equal masses of air and water would absorb the same number of cosmic rays. It was found later that, due to the peculiar nature of cosmic rays, more new ones would be created in the air layer than in water. While the numbers are close, more accurate readings would have disclosed a difference between the radiation levels at Arrowhead Lake and 6 feet down in Muir Lake. Section 6.6 discusses this phenomenon in more detail.

Among the main conclusions of Millikan were: *penetrating rays exist, they have a wide range of energies, and they come to earth with equal intensity, day or night, and at all hours of the day with nearly the same intensity in all directions.*

Since ionization was found beyond the equivalent of six feet of lead it was evident that this cosmic radiation was a good deal more energetic than even the hardest known gamma rays, a powerful form of photon radiation. To penetrate six feet of lead, charged particles would require an energy that was then considered impossibly large and much greater than possessed by gamma rays. Consequently, Millikan assumed, albeit incorrectly, that this cosmic radiation must consist of rays, since they have a lesser interaction with air in the atmosphere. One misguided admirer writing in the *Proceedings of the American Philosophical Society* dubbed the rays "Millikan Rays". But Millikan, realizing that they came from the cosmos, gave the name "*cosmic rays*" to this new radiation.

2.4 Final Confirmation

In the early 1930's the German physicist Erich Regener and his group perfected the technique of automatic recording electroscopes carried by balloons and sometimes carried to great depths in water. These highly accurate measurements of cosmic ray ionization versus height finally convinced most scientists that the cosmic radiation was of extraterrestrial origin. An example of one of Regener's charts is shown in the accompanying Figure 2.6. The atmospheric depth is on the horizontal axis in mass per unit area of the air layer above the electroscope. The vertical scale gives the number of ion pairs produced per cubic centimeter per second by cosmic rays at standard pressure and temperature. In these units the intensity at sea level is about 5 ion pairs per cc per second.

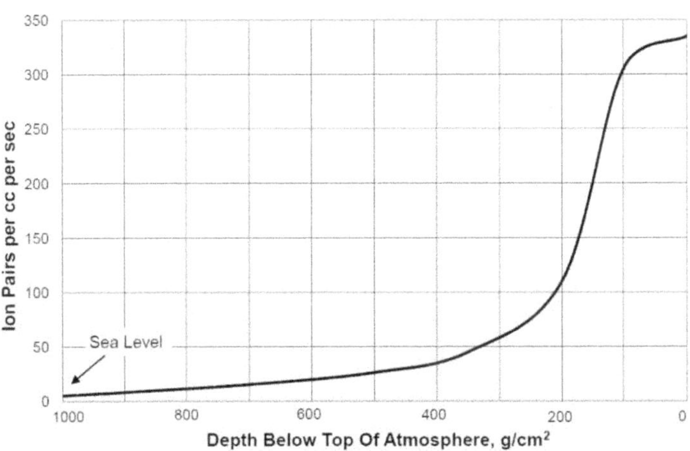

Figure 2.6 Erich Regener's very accurate ionization readings in the atmosphere.

Both Regener and Millikan believed cosmic radiation to be high-energy gamma rays. In fairness, we must note that at this time most physicists agreed with them that cosmic radiation was photonic in nature. This was probably because they could offer no other plausible explanation. (We know now that cosmic rays are, for the most part, charged particles and do indeed have enormous energies.) Nevertheless, Regener's and Millikan's work in confirming the source of the penetrating radiation added credence to Hess's and Kolhörster's work and satisfied the scientific community as to its origin.

After cosmic rays were confirmed to originate in outer space, only limited experiments were done to determine the type of radiation involved. This is because the theory and equipment of the period made it difficult to perform cosmic ray experiments. It was also because most physicists believed they had to be gamma rays, as no other particles of such energy were known.

With the corroboration of the existence of cosmic rays, scientists now knew that extraterrestrial sources were sending high-energy bullets into our atmosphere. And for some, as a matter of curiosity, it became necessary to investigate the nature of these bullets and to use them as probes to investigate matter in detail.

The realization that cosmic rays existed was instrumental in starting one of the greatest scientific quests in history; to wit, what are cosmic rays and how are they produced? This search also led to some of the greatest controversies between scientists, as we shall see in the next chapters.

Important particle physics studies began with cosmic rays and continue today. Through the intervening years many fundamental discoveries have been made concerning the nature of matter and the Universe. Perhaps, instead of cosmic rays, they should have been called *cosmic messengers*.

Chapter 3

3 Atomic Structure and Cosmic Rays

3.1 Studying Cosmic Rays – A Complicated Problem

Up until around 1927 most scientific efforts were concerned with determining the source of the mysterious ionization that was responsible for discharging a high quality electroscope. As noted in the previous chapter, that search was ultimately concluded in the early 1930's when most scientists agreed that *extraterrestrial* radiation was responsible. These cosmic rays were *assumed* to be high-energy gamma rays (photons) not charged particles as in our modern view - Figure 3.1. However it was some time until serious attempts were made to find out what cosmic rays actually were and how they reacted with the Earth.

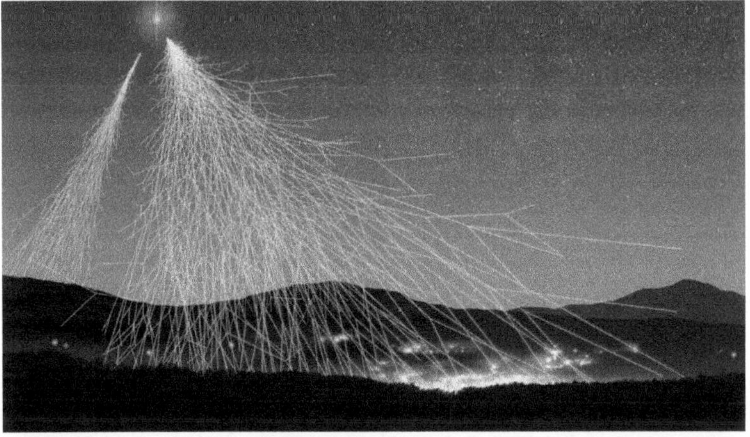

Figure 3.1 Modern view of cosmic rays impacting Earth and resulting air showers.

With the theory and equipment of this period it was very difficult to perform experiments on cosmic rays. At that time, methods used to study x-rays and radioactivity, were useful in a laboratory environment but cosmic rays were of much higher energy and seemed to come from everywhere. Also complicating the problem is that the percent of cosmic radiation on the Earth's surface is small compared to terrestrial radiation and therefore harder to measure.

In order to appreciate the situation scientists found themselves in, it will be helpful to review some of the theory and experimental methods of that time. The understanding of these subjects is intimately related to cosmic rays and occurred at about the same time as the study of cosmic rays began. In fact many developments in the fields of atomic structure, radiation and cosmic rays went hand in hand. Therefore in the interest of better understanding our subject we need to delve deeper into the physics of matter and its interaction with radiation.

3.2 Atomic Structure

First consider the concept of *matter*. Basically matter is any material that occupies space and has mass. It is essentially all the stuff around us. Matter is made of *elements*, with over 100 elements known. Some common examples are gases like hydrogen (H) and oxygen (O), and solids like gold (Au) and silver (Ag). Elements have a unique and identifiable atomic structure. A categorization of atoms into groups that share similar chemical properties is called the *Periodic Table of the Elements*. An atom is the smallest particle of an element that has the properties characterizing that element.

It is important to realize that the theory of atoms and atomic structure is based on experimental evidence and mathematical models. Atoms are difficult to observe directly even with the most powerful microscope, but their properties can be explored indirectly by modeling and laboratory studies. An early model of an atom is shown in Figure 3.2.

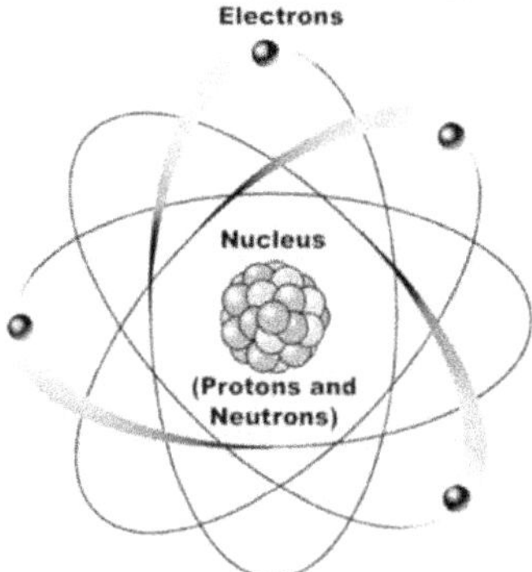

Figure 3.2 Early picture of an atom with a nucleus of protons and neutrons.

The central portion of an atom, the *nucleus*, contains smaller sub-atomic particles called *protons* and *neutrons* (except for common hydrogen which has only one proton). Protons are very dense, positively charged subatomic particles. Neutrons are dense, uncharged subatomic particles. Protons and neutrons have nearly the same mass and are built from even smaller particles called quarks.

Orbiting very rapidly about the nucleus of an atom are negatively charged particles called *electrons*, a type of lepton. It is generally assumed that electrons are moving so fast, that it is difficult to locate their position at any given moment. Consequently we tend to view electrons as a cloud of charged particles hovering about the nucleus. Electrons are very light compared to the proton and neutron having only about 0.06% of an atom's mass. Clearly most of the mass of an atom is concentrated in the nucleus.

As of now, the *Standard Model* in physics recognizes two types of fundamental particles; *matter* particles, some of which combine to produce the world about us, and *force* particles – one of which, the photon, is responsible for electromagnetic radiation. *Matter* particles are *fermions* while *force* particles are *bosons*.

Matter particles consist of six quarks and six leptons. These subatomic particles cannot be freed and studied in isolation. There are currently hundreds of identified particles made from combinations of these twelve fundamental particles and scientists are still finding more. Cosmic rays and particle accelerators are used for these investigations. The study of these numerous particles quickly gets very complicated, so we won't delve into all of them now. However, as we will see later, several of them like pions, kaons and their progeny are an important component of cosmic ray showers. Current research continues into the structure of the atom including the study of quarks, leptons and their associated force carriers such as photons, gluons and bosons.

Around the time of the discovery of cosmic rays, physicists were just learning that classical physics could not provide the answers about how electrons behaved in atoms. Sir Ernest Rutherford achieved one of the first breakthroughs in 1911 by establishing that the mass of the atom is concentrated in its nucleus. He also proposed that the nucleus has a positive charge and is surrounded by negatively charged electrons, which had been discovered in 1897 by J. J. Thomson. This seemed logical since if negative electrons were in an atom they would need to be balanced by positive charge in order for the atom to be electrically neutral.

Subsequently the existence of a positively charged particle in the nucleus was proved by Rutherford in 1919. He later named it the proton. The proton's charge is equal but opposite to the negative charge of the electron. Because the proton had such a large mass compared to the electron he suggested the atom had a miniature, dense nucleus surrounded by a cloud of nearly weightless electrons. There were a few problems with the model, however. According to classical physics, the electrons orbiting the nucleus should lose energy until they spiral down into the center, collapsing the atom.

A Danish scientist, Niels Bohr, while at Manchester University, adapted Rutherford's nuclear structure to Max Planck's quantum theory of 1901 and so obtained an improved planetary model of atomic structure (Figure 3.3). In this model the center of every atom has a nucleus, which is comparable to the sun in our Solar system.

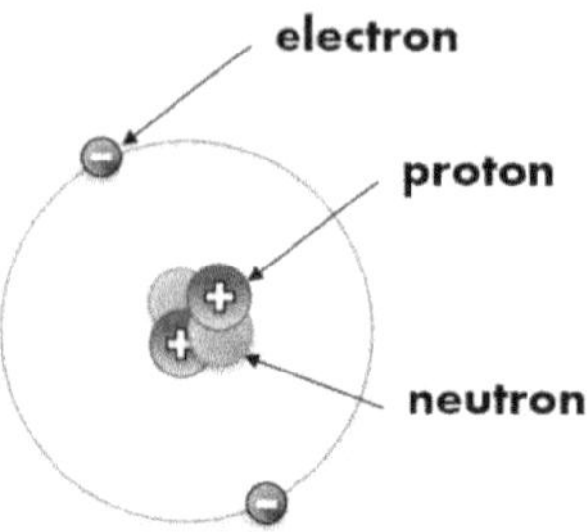

Figure 3.3 Illustration of Rutherford-Bohr model for Helium atom.

Electrons moved around the nucleus in "orbits" similar to the way planets move around the sun. That way, electrons existed at set levels of energy, that is, at fixed distances from the nucleus. If the atom absorbed energy, the electron jumped to a level farther from the nucleus; if it radiated energy, it fell to a level closer to the nucleus. His model was a huge leap forward in making theory fit the experimental evidence that other physicists had found over the years.

While scientists now know that atomic structure is more complex, the Rutherford-Bohr model is still a useful approximation to begin understanding about atomic structure, which, with later improvements remains valid to this day.

The neutron is the other type of matter particle found in the nucleus. A British physicist, Sir James Chadwick discovered it in 1932. The neutron carries no electrical charge and has about the same mass as the proton. With a lack of electrical charge, the neutron is not repelled by the cloud of electrons or by the nucleus, making it a useful tool for probing the structure of the atom.

3.3 Radiation and Ionization

Whenever the word radiation is mentioned it raises health concerns. But not all radiation is alike or harmful. When energy, in the form of energetic particles or energetic waves, moves through space it is said to be radiating. It is one of the most studied processes of physics. Because radiation radiates through space – its energy being conserved in vacuum – the intensity of all types of radiation follows an inverse square law of power with regard to distance from its source.

Radiation is commonly divided into two broad classes depending on the way it interacts with normal matter. The first kind is *ionizing radiation*, which comes from cosmic rays, radioactive materials and x-ray machines. Exposure to ionizing radiation can cause cellular and molecular changes such as mutations and chemical changes.

The second kind is *non-ionizing radiation,* usually electromagnetic radiation such as visible light, which comes from sources other than radioactive materials and x-ray machines. (Another way to think of electromagnetic radiation is as a stream of particles of energy called photons). The main difference between ionizing and non-ionizing radiation is the amount of energy it possesses. Non-ionizing radiation such as radio waves, microwaves, radar, and visible light is not able to break chemical bonds. Consequently this exposure is not related to events such as mutations or microscopic damage. It disperses its energy mainly through heat.

Ionization occurs when the radiation carries enough energy to remove an electron from an atom or molecule. It can be produced with *either* matter particles or photons. The most common ionizing matter particles include electrons, protons, and alpha particles. The most common photons with enough energy to ionize are short wavelength x-rays and gamma rays (γ-rays). Radioactive elements such as uranium, plutonium, cesium and cobalt emit high-energy ionizing radiation of both matter particles and photons.

Of the radiation known at about the time of the discovery of cosmic rays, those emitted by radioactive materials had the highest energies. Examples of these are the energies of alpha particles, beta particles and gamma rays, which are on the order of *millions* of electron volts. By contrast, the non-ionizing photons of visible light have energies only on the order of 10 electron volts. (An electron volt, abbreviated eV, is the kinetic energy of an electron accelerated by a potential difference of one volt.). Cosmic rays involve particles with energies of millions of electron volts (MeV) and sometimes higher to over one thousand million electron volts (GeV).

3.4 Interaction and Absorption of Radiation in Matter

Detection and characterization of radiation depends upon how it interacts with matter. When any kind of radiation penetrates matter some of the radiation may be *absorbed* completely, some may be *scattered* and some may pass straight through without any interaction at all. Absorption and scattering processes can be explained in terms of interactions between particles. To wit, particles in a beam of radiation strike particles in the material and are either stopped or scattered.

Two broad kinds of processes are known by which a particle traveling through matter can lose energy.

In one case the energy loss is gradual with the particle losing energy nearly continuously through many interactions with the surrounding material. In the second case the energy loss is *all or nothing*, with the particle moving without any interaction at all through the material until it loses all its energy in a single collision. For example, the motion of charged particles through matter is characterized by gradual energy loss whereas photon interactions are of the "hit or miss" type.

Charged particles lose their energy bit by bit through many ionization events. A beam of such particles all having the same mass, charge, and kinetic energy will slow down at nearly the same rate as they travel through matter. Thus they will all come to rest after covering the same distance. This distance is called the *range* and depends on the initial energy of the particles and material of the absorber.

When dealing with a beam of particles of the "hit or miss" type, like photons, a different process is involved. In this case a certain *fraction* of the initial quantity of photons traversing a given mass will be lost from the beam through interactions. This fraction depends on the thickness of the absorber, the material of the absorber, the initial energy of the particles in the beam, and the value of that energy. Theory and measurements have shown that the number of particles (radiation intensity) will decrease in an *exponential* fashion with the thickness of the absorber with the rate of decrease being controlled by a linear attenuation coefficient μ.

Expressing this formally we say that for thicknesses, from $x = 0$ to any other thickness x, the radiation intensity will decrease from I_0 to I_x, according to: $I_x = I_0 \exp(-\mu x)$ where x is the distance, exp is the exponential function and μ is the linear attenuation coefficient. Figure 3.4 shows radiation absorption curves (100 keV initial energy), for water (μ =0.167 per cm) and aluminum (μ=0.435 per cm).

The reciprocal of the linear attenuation coefficient μ has units of length and is often called the *mean free path.* The mean free path is the average distance a photon travels in the absorber before interacting. It is also the absorber thickness that produces a transmission of l/e, or 0.37 times the initial intensity. For 100 keV photons in water it is about 6 cm. For air, with $\mu = 0.000195$ per cm, the mean free path is about 5100 cm.

The thickness of absorbing material needed to reduce the incident radiation intensity by a factor of two is called the Half Value Layer (HVL). For water, as shown in Figure 3.4, the intensity is reduced to 50% at a thickness of 4.15 cm, For aluminum, the HVL is 1.59 cm and it is a more efficient absorber of 100 keV radiation.

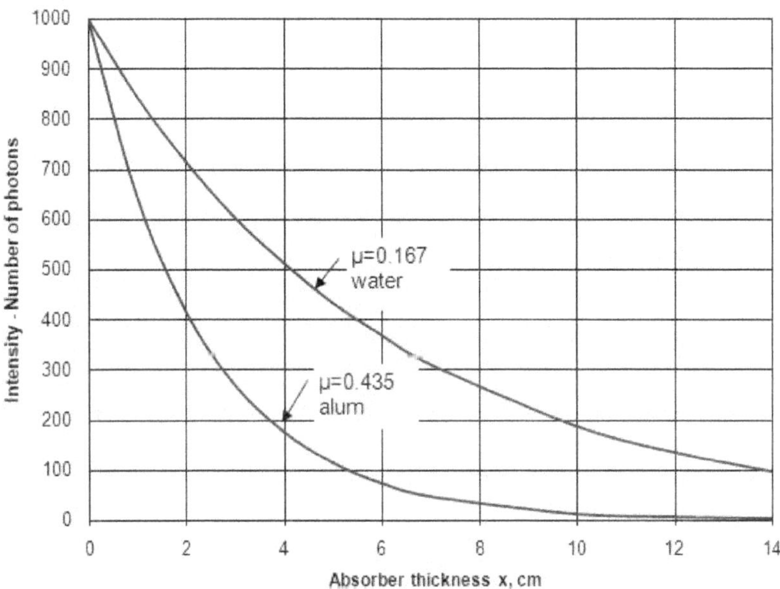

Figure 3.4 Photon radiation absorption curves for water and aluminum, 100 keV.

It is clear from the above that the fraction of photons absorbed depends on the material and thickness of the absorber. Hence, when comparing radiation absorbers it is necessary to consider density (mass per unit volume) and thickness. But because the thickness of an absorber can vary greatly, it has been customary to define an absorber not only by its thickness but by its mass per unit area – usually expressed in grams per square centimeter (gm/cm^2). Hence absorption curves are usually plotted with respect to this parameter.

Physicists in the 1920's used absorption curves extensively to measure and predict the energies of x-rays, gamma rays and cosmic rays in all types of materials. Absorption curves are still a valuable tool for studying radiation today.

Unlike photons, movements of charged particles are *direct* ionizing processes since coulomb interaction with matter directly causes ionization and excitation of atoms. The two most common charged particles are alpha (α) particles and beta (β) particles. An α particle consists of two protons and two neutrons bound together into a particle identical to a helium nucleus. Alpha particles are commonly emitted by all of the larger radioactive nuclei such as uranium, thorium, actinium, and radium.

Alpha particles interact strongly with the electrons of atoms, imparting enough momentum (due to the Coulomb interaction) to ionize the atom. In one interaction with an atom the alpha particle will lose only a small fraction of its energy. Many interactions are required for the alpha particle to lose all its energy. However, it interacts with many of the atoms along its path and loses all its energy in an unexpectedly short distance.

Beta particles are high-energy, high-speed electrons or positrons emitted by certain types of radioactive nuclei such as potassium-40. They are also known as beta rays. The interactions of charged particles are discussed in more detail in Section 3.5.

Indirect ionizing radiation can be caused by uncharged matter particles or photons. The two most common particles are neutrons and γ-rays (gamma rays). During interaction with matter they can transfer energy to charged particles, nuclei and atom electrons due to electromagnetic or nuclear interaction. After the energy is lost in the interaction, the radiation has reduced energy and may leave the material or interact further with it. The interactions of photonic particles are discussed in more detail in Section 3.6.

The accompanying Figure 3.5 shows the relative abilities of particles to penetrate solid matter. Alpha particles (α) emitted by radioactive nuclei are stopped by a sheet of paper, while beta particles (β) are stopped by an aluminum sheet. Gamma radiation (γ) is damped when it penetrates concrete or lead. Gamma rays being neutral will only lose energy when they hit something like an atom. Consequently when a gamma ray passes through matter, the probability for absorption is proportional to the thickness, density and material type. The total absorption of a group of gamma rays is an *exponential* decrease of their number with distance from the incident surface.

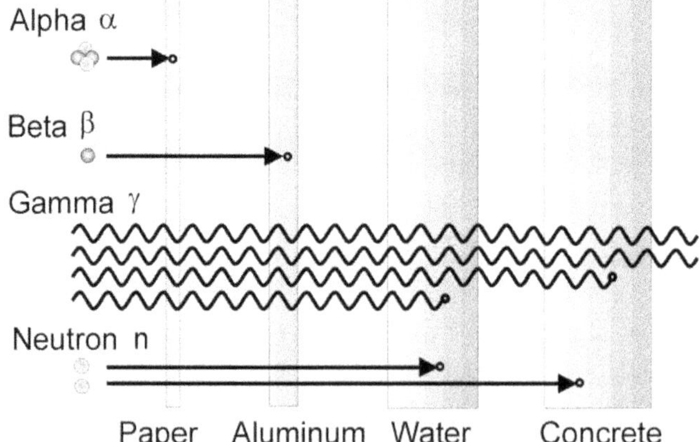

Figure 3.5 Relative ability of particles to penetrate matter.

High-energy neutrons are very penetrating and can travel hundreds of meters in air and several meters in common solids. They typically require hydrogen rich shielding, such as concrete or water, to block them within distances of less than a meter.

3.5 Charged Particle Interactions with Matter

Particles with charge such as electrons, protons and alpha particles, use the *electromagnetic interaction* or one of the two kinds of *nuclear interactions*, the *weak* interaction or the *strong* interaction. Electromagnetic interactions involving collisions with electrons in the absorbing material is by far the most common. This process is far different than neutral particles such as the neutron, which can interact only through the nuclear interactions. Thus charged particles can be detected directly by their electromagnetic interactions whereas neutral particles have to use nuclear interactions, which in-turn produce charged particles before their presence can be detected.

Electromagnetic interactions consist mainly of two mechanisms: (a) excitation and ionization of atoms, and (b) bremsstrahlung, the emission of photonic radiation when a charged particle is severely accelerated, usually by interaction with a nucleus. (There is a third kind of interaction, producing Cherenkov radiation, which plays an important role in the detection of *very high-energy* charged particles but absorbs only a small amount of energy.) The contribution of each mechanism depends on the charge, mass and speed of the incident particle as well as the atomic numbers of the elements, which make up the absorbing material.

Each charged particle may suffer many interactions along its path before it finally comes to rest, but only a small fraction of its energy is lost at each interaction. For example, a typical alpha particle might make thousands of collisions before it stops. Hence the energy loss can usually be considered as a continuous process. Charged particles are deflected or scattered at each interaction. Although the amount of scattering at each collision may be small, the cumulative effect may cause a large change in the direction of travel. Occasionally an incident particle will pass very near a nucleus and then there will be a single large deflection. This nuclear scattering effect is most pronounced for light incident particles interacting with heavy target nuclei.

In the early 1900's it was common practice to study radiation by its property of ionizing gases. The manner in which different types of radiation ionize gases was thought to be well understood. This impacted the study of cosmic rays since, until more modern methods were developed, the only way to study cosmic radiation was through its ionization effects on gases.

When a high-energy charged particle like an electron (sometimes called a beta particle) travels through a gas, its electric force may excite or disturb the electrons in the gas molecules. With a sufficiently violent disturbance an electron from the molecule may break loose and leave behind a positively charged molecule, or positive ion. The freed electron will remain free for a short time and wander around the gas until it attaches itself to another atom. But the initial charged particle, having lost some energy, continues the process leaving behind in its wake a trail of ions, until it too finally comes to rest.

This trail of ions can be made visible through the use of an ingenious instrument known as the *cloud chamber* This instrument was invented around 1911 by the Scottish physicist C. T. R. Wilson and is also known as a Wilson cloud chamber, or expansion chamber, to distinguish it from later types (Figure 3.6). The cloud chamber has, as the detecting material, a supersaturated vapor that condenses to tiny liquid droplets around ions produced by the passage of energetic charged particles, such as alpha particles, beta particles, or protons. In a Wilson cloud chamber, super-saturation is caused by the cooling induced by a sudden expansion of the saturated vapor by the motion of a piston or an elastic membrane, a process that must be repeated with each use.

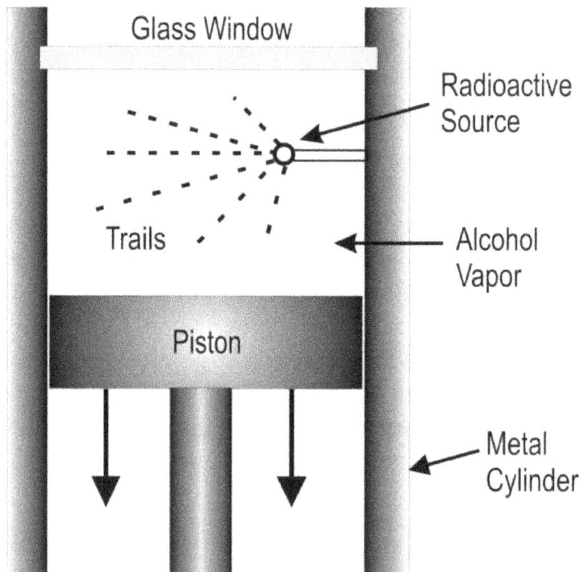

Figure 3.6 Wilson expansion type cloud chamber. Trails viewed thru top window.

More specifically, the chamber contains a mixture of alcohol vapor and either air or some gas such as argon. If one suddenly enlarges the enclosed space by pulling back on the piston, the gases expand and the temperature drops. If the temperature drop is great enough, the alcohol vapor condenses into droplets, which form a fog in the chamber. Appropriate control of the expansion, however, will produce a somewhat smaller temperature change, and the vapor will condense only on the ion trails. Consequently, a charged particle traveling through the chamber will leave a track in the form of tiny visible drops that condense along the particle's ionization trail as shown in the accompanying Figure 3.7.

To be detected, the particle must be traveling through the chamber during its expansion phase. It if enters the chamber too soon the ion trail will diffuse away before the gases have cooled. If it enters too late the gases will warm up before the ion trail is formed. So, there is a critical sensitive period for detection of ion trails. Later cloud chamber designs have been much improved over the early Wilson chamber, but it was certainly a remarkable device that made the study of atomic structure possible.

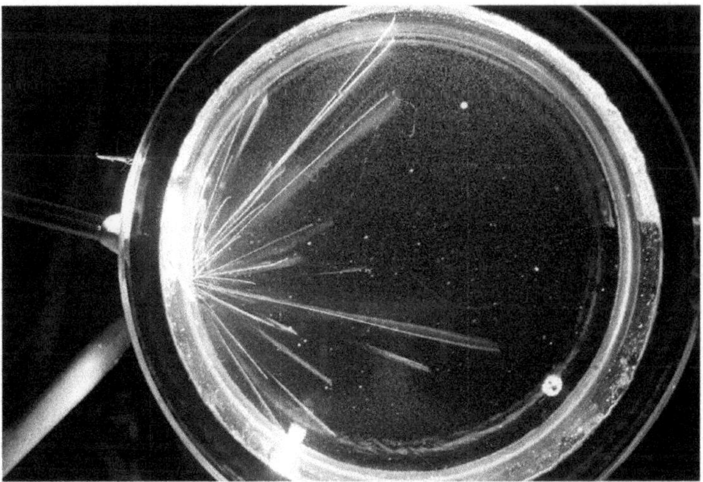

Figure 3.7 Replica of original Wilson cloud chamber showing ionization trails.

Fast moving particles with different electric charges and velocities produce ion trails of different densities in the gases. At a fixed speed the number of ions per unit length in the ion trail, increases with charge. This happens because the electric forces on the electrons of a molecule caused by a charged particle passing close by are proportional to the charge of the particle. Hence, particles of larger charge, such as alpha particles cause more violent disturbances than particles of smaller charge and ionize more molecules over the same distance.

Now, if the charge stays fixed, like an electron or proton, the ion density *decreases* with increasing velocity. This happens because the charged particle disturbs the molecule only when it is within a certain minimum distance. Hence the time interaction is longer for a slow moving particle than a faster one. This longer period of interaction provides a more effective force to disrupt the molecule. Therefore, slower particles ionize more densely than faster ones. These characteristics allow scientists to deduce much about a particle from its ionization trail. This technique is particularly effective when studying cosmic rays.

When the kinetic energy of a charged particle increases continuously its velocity eventually approaches the speed of light. The density track left by such a particle at first decreases with increasing energy, but then tends to a nearly constant value similar to a particle moving at nearly the velocity of light. Thus there is a lower limit on the amount of ionization produced by any charged particle. See Figure 3.8. An electron (or proton) moving at nearly the speed of light leaves an ion trail of the smallest possible density. They are called *minimum ionizing particles* and are the hardest to detect in a cloud chamber.

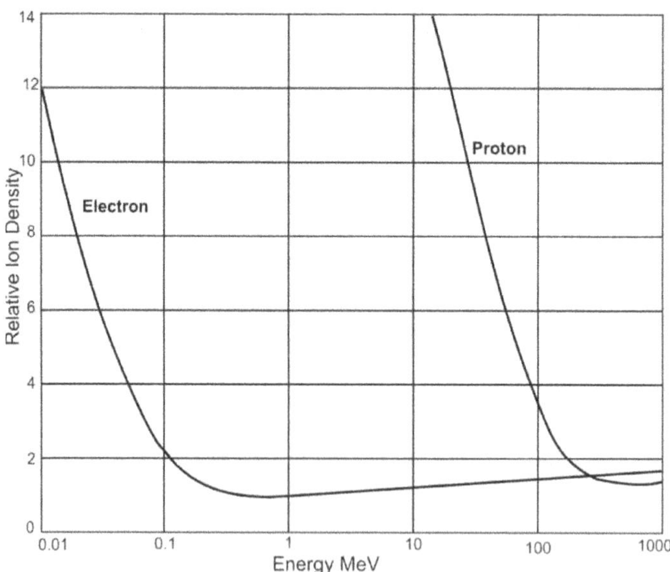

Figure 3.8 Relative ion density along the track of charged particle.

To illustrate, a high-speed electron, emitted from radioactive material, may have a kinetic energy on the order of 1 MeV, travels at about 94 percent of the speed of light, and behaves like a minimum ionizing particle. In air it produces about 50 ion pairs per centimeter. On the other hand, alpha particles from radioactive polonium have energy of 5.3 MeV and travel at only 5.4 percent of the speed of light. They produce about 24,000 ion pairs per centimeter in air. Consequently alpha particles are very easy to see in a cloud chamber while high-speed electrons leave a very thin trail and are harder to see.

From the point of view of scientists in the 1920's studying cosmic rays, it was commonly believed that all charged particles traveling through matter lost all their energy by electromagnetic interactions (ionization). As we will see later on, this belief turned out to be incorrect.

3.6 Photonic Interactions in Matter

Gamma rays, x-rays and light are all photons, but with different energies. The energy of a photon is inversely proportional to its wavelength (λ). In other words, its energy increases with decreasing wavelength. For example, gamma rays are photons with very short wavelength and high energy. They are emitted from an unstable atomic nucleus and have high penetrating power.

Photons can produce ionization in materials, but by different processes and much more rarely than those involving charged particles. To illustrate this difference, consider a 2 MeV electron and a 2 MeV photon in air. While they both have the same energy, the electron travels about 0.0002 meters before ionizing a molecule while the photon travels an average distance of 170 meters! But unlike the electron, which loses only a fraction of its energy at the encounter, the photon loses most or all of its energy. Such photon interactions are often described as "hit or miss".

Photons interact with matter in a variety of ways, depending on their energy and the nature of the material encountered. Three interaction mechanisms are the photoelectric effect, Compton scattering, and pair production.

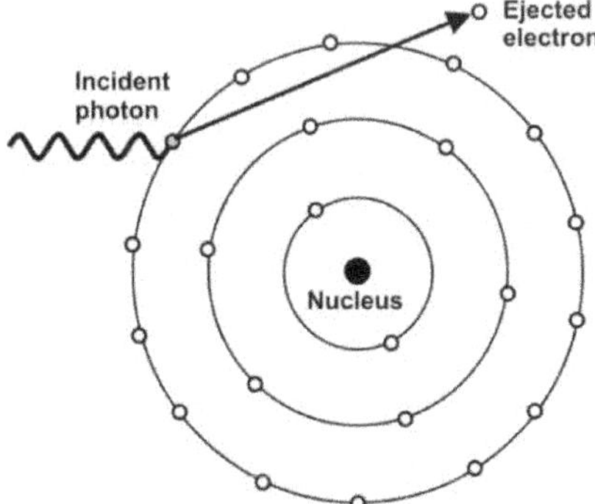

Figure 3.9 Photoelectric effect, usually occurs with photon below 1 MeV.

The *photoelectric effect* occurs when a low energy photon strikes an orbital electron, as shown in Figure 3.9. The total energy of the photon is expended in ejecting the electron from its orbit. The result is ionization of the atom and expulsion of a high-energy electron. The photoelectric effect is most predominant with low energy photons and rarely occurs with photons having energy above 1 MeV.

Compton scattering is an elastic collision between an electron and a photon, as shown in Figure 3.10. In this case, the photon has more energy than is required to eject the electron from orbit, or it cannot give up all of its energy in a collision with a free electron. Since all of the energy from the photon cannot be transferred, the photon must be scattered. The scattered photon must have less energy, or equivalently a longer wavelength. The result is ionization of the atom, a high-energy electron, and a photon at a lower energy level than the original. Compton scattering is most predominant with photons at an energy level in the 1.0 to 2.0 MeV range

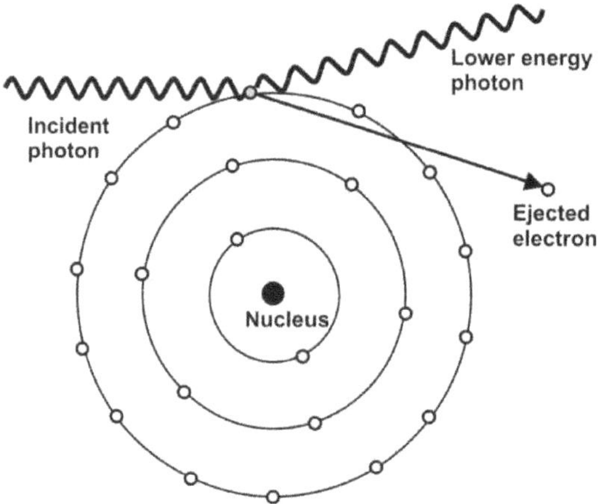

Figure 3.10 Compton Effect, photon ejects electron and lower energy photon.

At higher energy levels, *pair production* predominates other effects. When a high-energy photon passes near enough to a heavy nucleus, the photon disappears, and its energy reappears in the form of an *electron* and a *positron* as shown in Figure 3.11. (A positron has the same mass as an electron, but has a positive charge.) This transformation of energy into mass must take place near a particle, such as a nucleus, to conserve momentum. The kinetic energy of the recoiling nucleus is very small so nearly all of the photon's energy in excess of that needed to supply the mass of the pair appears as kinetic energy of the pair. For this reaction to take place, the original photon must have energy of at least 1.02 MeV.

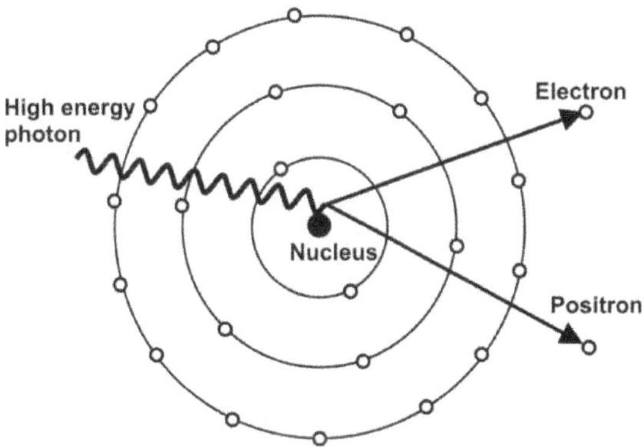

Figure 3.11 Pair production by high-energy photon near heavy nucleus.

Subsequently, the electron loses energy by ionization. The positron that is formed quickly disappears by reconversion into photons in the process of annihilation with another electron in matter. The probability of pair production increases significantly for higher energy photons. Patrick Blackett, in 1933, was the first to discover pair production using cosmic rays and a cloud chamber.

Scientists of the 1920's knew about the photoelectric effect and Compton scattering. But the theory of pair production was not discovered yet. Since the photoelectric effect was only relevant for lower energy particles it was thought that high-energy photons could only lose energy by Compton collisions. These scientists did not realize that high-energy mass particles and photons are absorbed more readily than was supposed.

3.7 Are Cosmic Rays Gamma Rays?

Because photons were the most penetrating radiation known at the time of the discovery of cosmic rays, most scientists in this infant field were convinced that the rays were simply photons of greater energy than previously studied. The reason for this is understandable – gamma ray photons traveled farther than any other known radiation. As we have seen in Section 3.3, relatively low energy gamma rays of 100 keV have a mean free path of over 5000 cm in air. But beta particles have a range of only a few cm in air and alpha particles even less. The most penetrating radiation of the time consisted of gamma rays from natural radioactive substances. But even this radiation was not as penetrating as cosmic rays. If cosmic rays were actually high-energy gamma rays, scientists expected their mean free path to be enormously higher, but there was no experimental evidence for this assumption.

As more scientists began studying cosmic rays it became clear that the assumption that the rays were gamma rays had to be tested. Because more sophisticated instruments like the cloud chamber and Geiger tubes began to be used in experiments, questions started to emerge about the nature of this radiation from space. So, in the late 1920's increasingly complex theories were developed and experiments were carried out. After another decade or so of research, the real picture of cosmic rays began to emerge and it was nothing like the original assumptions.

Chapter 4

4 Early Theories, Developments and Conflicts

4.1 The Gamma Ray Assumption

Most scientists in the 1920's believed cosmic rays were enormously powerful gamma rays because they seemed to have greater energies than any other known particles and could travel very large distances. At this time the only known radiation came from radioactive substances in the form of alpha particles, beta particles (electrons) and gamma rays (photons). Of this group, gamma rays had by far the greater penetrating power. It was known that some gamma rays could travel hundreds of meters in air while for charged particles like high-speed electrons (beta rays) it was only a meter or less. So, it was natural for physicists to *assume* that cosmic rays were gamma rays but of much higher energy.

Bolstering this idea were theoretical predictions that higher energy photons would have a much larger mean free path. These results were based on the prevailing theory of the time, the Compton Effect (Section 3.5), which predicted that photon collisions in matter would have an increasing mean free path with increased photon energy. It had also been found in the laboratory that the more powerful gamma rays produced the greatest penetration in materials within the bounds of energies used.

Outside the laboratory it had been found, by the absorption measurements in air and water by Hess, Kolhörster, Millikan, Regener and others, that cosmic radiation was more penetrating than any other known type. But this radiation could not be tested in the laboratory using Compton's theory. So, if theoretical predictions were attempted, based only on the low energy laboratory data, they would require a huge extrapolation of the known absorption curves. Nevertheless, even in the face of enormous extrapolations, on the order of 10,000, some physicists were willing to estimate the energies of the theoretical cosmic rays from the shape of the lower energy absorption curves. As one physicist of the day, W. Swann, remarked, it is the equivalent of drawing a curve on a piece of paper one-foot square and then extrapolating for a distance of two miles for the purpose of drawing conclusions.

Thus, based on these extrapolations, it was concluded that cosmic rays were photons of energies ranging from 20 MeV to 220 MeV. This led to some great speculations, the most astounding one put forward by Millikan in 1928. Millikan developed a complex theory that cosmic rays were the result of atom building of heavier elements from the elemental hydrogen gas that filled the universe. This theory and other amazing developments involving cloud chambers and Geiger tubes are discussed in the next sections.

4.2 Millikan's Theory of Cosmic Ray Formation

Around the time of the lake experiments with Cameron (Section 2.2), Millikan developed several guiding principles for the study of cosmic rays. First, he concluded that cosmic particles were gamma rays entering the atmosphere isotropically from space. And secondly, he believed the gamma rays energy absorption properties could be established in much the same way as x-ray absorption properties.

Thus, it is not surprising that much of Millikan's theory critically depends on determining energy absorption curves of cosmic rays. Absorption curves are *exponential* in nature and characterized by a linear attenuation coefficient μ, which is the inverse of the mean free path length (Section 3.3). Determination of absorption curves assumes a parallel beam of constant energy photons impinging on the absorber. The coefficient μ is dependent on the absorber material and photon energy, and characterizes the slope of the curve.

Cosmic rays presented certain problems in finding an absorption curve. First of all it was clear that cosmic rays did not form a single vertical beam of photons but instead emanated from the atmosphere in all directions. To some extent this effect can be corrected, albeit with some effort, through the use of a mathematical model. A second problem was that it was not known if the radiation was all of the same energy. However, this fact might be established by plotting the experimental data to see if it fit an exponential curve. The μ thus found together with the theory of Compton scattering would allow the energy of the photons to be found.

Millikan carefully plotted the ionization rates of his electroscope with respect to absorber thickness and tried to fit the data to an exponential curve. He was frustrated because a single value of μ did not fit the curve. In a moment of insight, he realized that it could be represented by the *addition* of three exponential curves, each with a separate value of μ. Using the current theory of the Compton Effect, due to Dirac, Millikan calculated energies of 26 MeV, 110 MeV and 220 MeV for the three respective curves. Hence he concluded that cosmic radiation was a *combination* of those three energies.

While pondering the origin of these cosmic photons, Millikan was drawn to the idea that interstellar space was filled with hydrogen gas and perhaps those three groups of photons were a result of a process known as *fusion*. Fusion reactions are high-energy reactions in which two lighter atomic nuclei fuse to form a heavier nucleus. When they combine, some of the mass is converted into energy in accordance with the famous equation of Einstein, $E = mc^2$.

A helium atom weighs slightly less than four hydrogen atoms. Millikan reasoned that if the helium atom were formed from the four hydrogen atoms the excess mass would be given off by a gamma ray. The energy of this gamma ray would be 27 MeV and for practical purposes, close enough to the first group of photons. Hence Millikan concluded that the first group of photons resulted from the synthesis of helium in interstellar space.

Likewise a nitrogen atom consists of 14 hydrogen atoms and the amount of energy released would be about 100 MeV. Similarly, the building of oxygen from 16 atoms of hydrogen produces 120 MeV. So the energy of the second group of photons could be thought of as resulting from the synthesis of these atoms. Finally, if 28 hydrogen atoms are fused to form silicon the resulting energy released corresponds to the third group of photons.

As a final convincing argument, Millikan and Cameron compared the abundance of these elements on Earth to those they believed to be in space and found excellent agreement. Millikan therefore concluded that cosmic rays were the "birth cry" of atoms continuously created in space.

Although there were some criticisms from fellow scientists such as J. Robert Oppenheimer, the theory was accepted for several years with Millikan proclaiming that the creation of elements was ongoing throughout the universe. But new discoveries using cloud chambers and Geiger tubes were starting to cast doubt on the theory and cracks started to emerge.

4.3 Cloud Chamber Sees a Cosmic Ray

Cloud chambers (Section 3.4) are used to detect ionization trails of *charged* particles in a supersaturated space. Quite a lot of information about the particle can be deduced from studying the density (ions per cm) of the trail. But even more information, like the energy of the particle, can be found by placing the chamber in a strong *magnetic field*. The magnetic field causes the charged particle to bend by an amount proportional to the strength of the field, its velocity and the degree of charge. By analyzing the curvature of these tracks a wealth of information about the particle can be gleaned. This is very useful for studying alpha and beta particles emitted from radioactive material. However, in the case of gamma rays we are dealing with *uncharged* particles. So how are they detected?

Around 1900, C. T. R. Wilson, the inventor of the cloud chamber, placed an x-ray tube close to one of his early chambers. He saw minute streaks and patches of clouds throughout the region in the path of the beam. The clouds were mainly short thread-like objects, a few of them moving in straight lines and some of them even looping round. Wilson saw this as a direct proof that x-rays liberated energetic electrons in the gas, and that these electrons caused ionization along the path of the x-rays. But Wilson could not explain theoretically why this happened. A. H. Compton observed this effect in 1923 and ultimately provided the reasoning that resolved this enigma. Later in 1923 Wilson, using a perfected chamber, took photographs of the tracks of fast electrons produced by x-rays, explicitly acknowledging Compton's work.

The explanation of the Compton Effect is given in Section 3.4 where we saw how photons can have collisions with electrons. In turn, these high-speed "knock off" electrons can create ionization trails in a cloud chamber like those seen by Wilson. Hence x-rays or gamma rays with sufficient energy, although not direct ionizers, may be detected in a cloud chamber by the recoil electrons they generate.

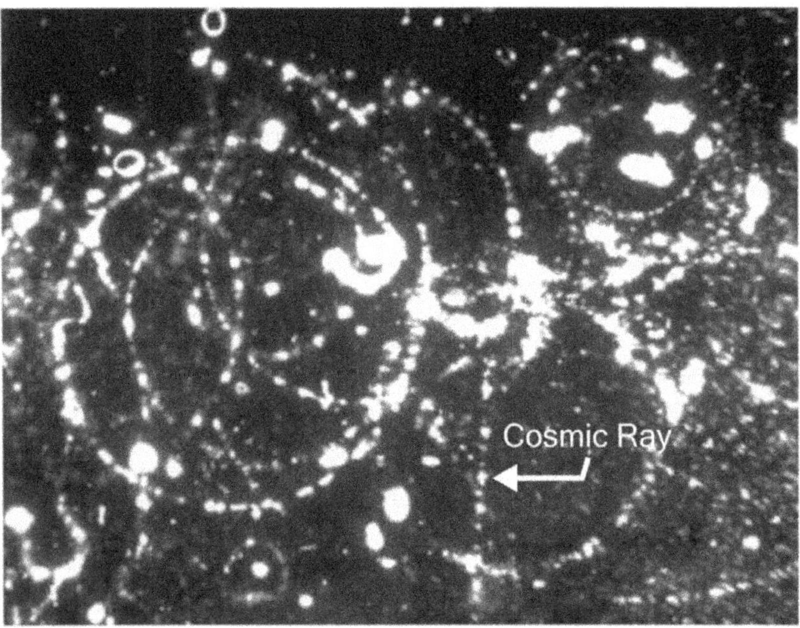

Figure 4.1 Cloud chamber photo of a possible cosmic ray obtained by Skobeltzyn in 1928. It can be seen as the downward line of dots near the arrow.

In Leningrad in 1928, a young scientist, Dimitrij Skobeltzyn, made an amazing discovery while examining gamma rays emitted by radium. He had set up a small Wilson cloud chamber between the poles of powerful magnet of 1500 gauss. As part of the experiment he would routinely examine the photographs of the curved or circular tracks of the recoil beta particles caused by the gamma rays and make calculations of the energies of the particles involved. While studying one specific photograph he noticed a particular track that did not show much curvature. A copy of that photo is shown in Figure 4.1. He assumed that the nearly straight track must be from a high-speed electron. But when he calculated its energy, it was found to be in excess of 15 MeV, a value much higher than that associated with radioactive materials.

Skobeltzyn was not sure at the time if this was a cosmic ray. But in examining an additional 613 photos he found 32 more nearly straight tracks that emanated from outside the chamber. All of the 32 tracks had energies exceeding 15 MeV. Noting that their rate of occurrence would roughly account for the amount of ionization attributed to cosmic rays, he correctly ascribed them to the penetrating cosmic radiation.

More specifically, he interpreted the particles as being Compton recoil electrons generated by the high-energy gamma radiation coming down from above. This seemed to endorse the gamma ray assumption that cosmic rays were high-energy rays. But, on the other hand, it also showed that a very high-energy matter particle, like an electron, could be generated with enormous energy.

4.4 Geiger Counters Transform Cosmic Ray Studies.

A great new invention was born in 1928 that would revolutionize the study of cosmic rays. This device, the Geiger-Müller counter, became one of the most widely used tools of experimental physics in the last century. Because of its simplicity, low cost and ease of operation this device has continued to be used up to the present time.

It had its beginnings at the University of Manchester in 1912, when Hans Geiger, while working in the laboratory of Ernest Rutherford, developed a revolutionary electrical counting device known as the *point counter*. Prior to this time the optical method was used, which consisted of laboriously counting tiny flashes of light on a scintillation screen. This new method relied only on the electrical effects of particles and ions.

This unusual device (Figure 4.2) consisted of a thin needle shaped rod fitted into a metal box. It was connected to a battery (on the order of 1000 volts), through a resistor R, which maintains the rod at a positive voltage with respect to the case. There was a thin window on the case that would permit ionizing particles to enter and also allow different gases to be used in the box.

When ionizing particles entered the box, a discharge would occur and the voltage on the rod would suddenly drop in pulse fashion. An electroscope (Section 1.2) connected to the rod could easily witness this voltage drop. After the discharge, the rod would again charge to the positive voltage through the resistor and be ready for the next particle.

The discharge happened because the radiation introduced into the chamber caused ionization, with the ions being drawn to the needle or case depending on their polarity. But because of the high voltage on the rod, the ions became accelerated, creating many new ions and eventually causing an avalanche of ions in the gas. This caused the gas to become very conductive and an electrical discharge would occur.

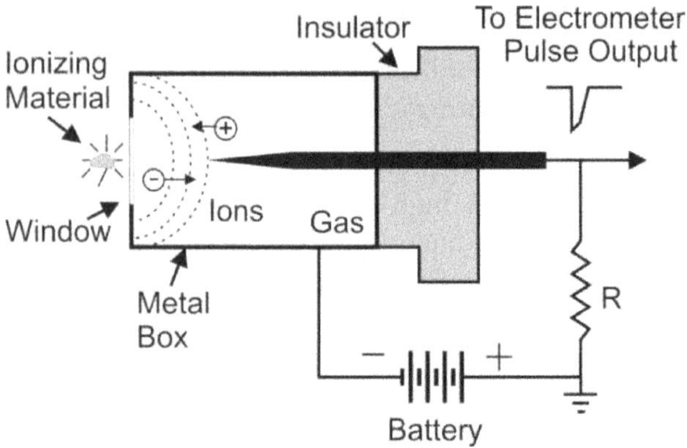

Figure 4.2 Geiger's revolutionary point counter with pulse output.

This device did work, after a fashion, but it was unreliable, sporadic and only seemed to work in the vicinity of the point, which was too small to be of use in cosmic ray studies. But it was used in an important radiation study that verified the Compton Effect (Section 4.4). The outbreak of World War I interrupted further work on this device, but Geiger was determined to improve on this concept.

Eventually in 1928, working with his PhD student W. Müller at Kiel University, Geiger perfected a greatly improved counter. The *Geiger-Müller counter* (G-M counter) was similar to the original point counter in that it used the avalanche effect, in which a single ion pair produced in the gas triggers a discharge. However its design was a great deal different and did not rely on a needle shaped rod.

The G-M counter consists of a gas filled cylindrical metal tube with a very thin wire stretched along its axis and insulated from the tube (Figure 4.3). A large voltage (500 to 1000 volts) is applied to the central wire, with respect to the metal tube, through a resistor R. This creates a strong electric field between the wire and tube. If any ions are produced in the space between the wire and tube they will be accelerated and cause an avalanche and thereby a discharge, resulting in a negative going pulse output.

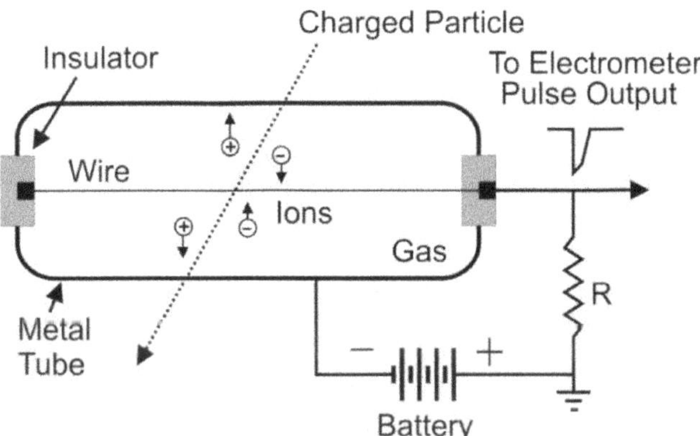

Figure 4.3 Geiger-Müller counter being triggered by cosmic ray.

As the working of G-M counters is based on gas multiplication the fill gas must be free of any traces of electronegative impurities like oxygen or water vapor. Noble gases like argon or helium are popular choices for the fill gas. A second component is added to the fill gas for quenching the discharge, so multiple spurious discharges do not occur. Ethyl alcohol in a concentration of 5% is often used for this purpose. And finally the pressure of the gas is lowered to a tenth of an atmosphere to reduce the operating voltage.

It was the *sensitivity* of the device that caught the attention of the physics community. In their early publications Geiger and Müller emphasized the capability of the device to respond to even the weakest radiation. This meant that it would substantially reduce the time required to perform lengthy experiments. To promote the device, Geiger invited notable scientists to his laboratory to observe the counter in action. Using a loudspeaker they impressed their observers with the clicks of the new device and conveyed the sensation of the, hitherto unknown, perception of radiation. Thus the new G-M counter became an instrument for the measurement of radiation as well as for public demonstration.

One notable public demonstration was performed at the opening of the 1939 New York World's Fair. It was the first exposition to be based on the future, with an opening slogan "Dawn of a New Day". Albert Einstein (Figure 4.4) abandoned his own best judgment and participated in elaborate opening ceremonies staged to illuminate the fair by counting ten cosmic rays. The entire affair proved to be a comedy of errors.

First, due to a faulty sound system and his thick German accent, only a few words of Einstein's five-minute talk on cosmic rays could be heard. At the talk's end, as planned, the apparatus designed by the Bartol Research Foundation using three Geiger counters to detect and count cosmic rays, was enabled (Figure 4.5).

Figure 4.4 Albert Einstein and wife at New York World's Fair in 1939.

Numerous ringing bells and flashing lights signaled the capture of each ray. But when the tenth cosmic ray was captured and the switch thrown to illuminate the huge light on top of the spire-shaped structure known as the Trylon, the electrical system overloaded, which caused a major power failure. Thus ended this auspicious demonstration of Geiger counters and cosmic rays!

Figure 4.5 Cosmic Ray Machine for 1939 World's Fair. Dr. W. F. G. Swann, director of the Bartol Research Foundation of Franklin Institute, Philadelphia, is shown examining the triple coincidence Geiger counter made at the institute.

None the less, for real cosmic ray measurement applications the G-M counter is particularly appealing. Compared to the earlier point counter, the G-M counter is easier to build, is much more reliable and can be made in sizes up to several inches in diameter and several feet long. As discussed in the next section, one of the most famous experiments with cosmic rays involved the use of G-M counters.

4.5 The Legendary Coincidence Experiments

Cosmic ray research changed dramatically with the advent of the Geiger-Müller counter (G-M counter) in 1928. For the first time the physical nature of cosmic rays could be studied. Used alone as cosmic rays detectors, they did not have significant advantages over ionization chambers. But when used in *coincidence* arrangements they became a most powerful tool for cosmic ray experiments.

The coincidence technique had its beginnings in 1924 when it was used to study the newly discovered Compton Effect (Section 3.5), which was somewhat controversial at the time. It was not known if the effect was purely statistical or if it could actually be interpreted as a collision between a photon and an electron, with the system still observing the laws of conservation of energy and momentum. So, Walter Bothe and Hans Geiger devised an experiment that used x-rays and employed the early Geiger needle counters (Section 4.3) to detect the coincidence between the recoil electron and the scattered photon.

Two separate needle counters were arranged to monitor *simultaneously* the counts from the electron and photon. Their experimental results showed that the recoil electron and the scattered photon were coincident in time, to within 1 milli-second, confirming that the two events happened simultaneously as necessary for the Compton Effect. This was a major step in disproving the statistical hypothesis. But while the result was notable, it was not complete since it did not validate the conservation laws. (Soon afterward Compton and Simon using a cloud chamber proved that energy and momentum are conserved). Nonetheless, this coincidence experiment was a landmark in physics and an achievement for needle counters, but this may have been the last time they were used in a major study because of their finicky nature and the development of better devices.

At the July 7, 1928 German Physical Society meeting, the next generation counting device, invented by Hans Geiger and his pupil William Müller, was announced; the G-M counter (Section 4.4). According to them it was very responsive to external radiation and they strongly suspected that some of its spontaneous counts should be attributed to cosmic radiation.

Several weeks later, at the end of July of 1928, at an informal conference organized in Cambridge U.K., Dmitry Skobeltzyn, who was mentioned in Section 4.2, presented a talk on "The Intensities of Gamma Rays" and showed some photographs of what he thought were cosmic rays in his cloud chamber. Hans Geiger was in attendance and, following Skobeltzyn's remarks, announced that Werner Kolhörster and Walter Bothe (who were working in the same institute as Geiger) had devised a new method to register cosmic rays by the coincidence method of pulses in two G-M counters. It was here that physicists first learned of a new kind of counter experiment that would revolutionize cosmic ray research.

In their experiments in 1928, Bothe and Kolhörster set up two G-M counters, each with its own electroscope, for observing cosmic rays. They experimented with several arrangements of the counters and noticed when one was placed a small distance above the other, *simultaneous* deflections of the electroscopes often occurred. Such an arrangement is shown in Figure 4.6.

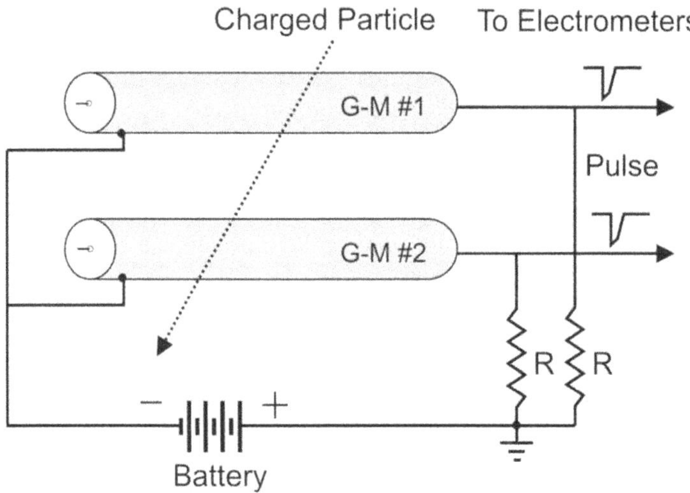

Figure 4.6 Coincidence arrangement using two G-M counters placed above each other. Simultaneous deflection of electrometers indicates passage of charged particle.

Upon reflection they realized that these coincidences could not be due entirely to chance since they became less frequent as the distance between counters was increased. They also found that the number of coincidences varied with direction with about three times more coincidences along the vertical with respect to the horizontal direction, exactly what you would expect if the radiation was coming from above.

Recall that charged particles are needed to trigger a G-M counter (Section 4.3). If the penetrating radiation were photons then Compton collisions would be required to produce the recoil electrons needed. But the probability of a Compton collision in the wall or gas of a counter is very small for a single counter, being on the order of 5%. But, assuming that this did cause a discharge in one counter, the probability of two Compton collisions caused by a single photon becomes extremely low, on the order of 0.25%. Hence, Bothe and Kolhörster correctly concluded that the observed coincidences must be due to the passage of single *charged* particles through both counters. Moreover, the particles were not ordinary beta particles from terrestrial sources since the counter walls were 1 mm thick and made of zinc, which would stop all such particles.

By itself this result was not at variance with the accepted view that the primary radiation falling on the atmosphere from space consisted of high-energy photons. Obviously, photons could undergo Compton collisions high in the atmosphere and produce ionizing recoil electrons. As was noted earlier it was believed that cosmic rays were photons of energies of 20 MeV to 220 MeV. Thus the recoil electrons produced from these photons would have more than enough energy to pass through the counter walls.

To examine this possibility Bothe and Kolhörster built a thick housing of iron and lead in which to place two G-M counters. This shield would block local terrestrial radiation from all sides and included a space between the counters to place absorbers. The initial experiment was made on the first floor of their building. In their paper Bothe and Kolhörster said; "*It soon stood out that the radiation which caused the coincidences must be extraordinarily penetrating, as in a short series of experiments the insertion of 1 cm and even 4 cm of lead as absorber gave no distinct decrease in the number of coincidences*".

Because of the limited dimensions of their apparatus, thicker absorbers could not be used. So for the final experiment they used a 4.1 cm thick block of *pure gold*, which was placed at their disposal for a short time. Gold has a higher density than lead and hence a greater ability to stop fast electrons. Now, with this arrangement (Figure 4.7), only particles with a range greater than 4.1 cm of gold could pass through both counters.

Compton electrons have low penetrating power and should be completely absorbed by even a very thin layer of gold. So the experimenters expected to see only a very small number of coincidences. To their great surprise they found that the rate of coincidences was still 76% of what it had been without the gold block. In other words, 76% of the charged particles present in the cosmic radiation at sea level could pass through 4.1 cm of gold!

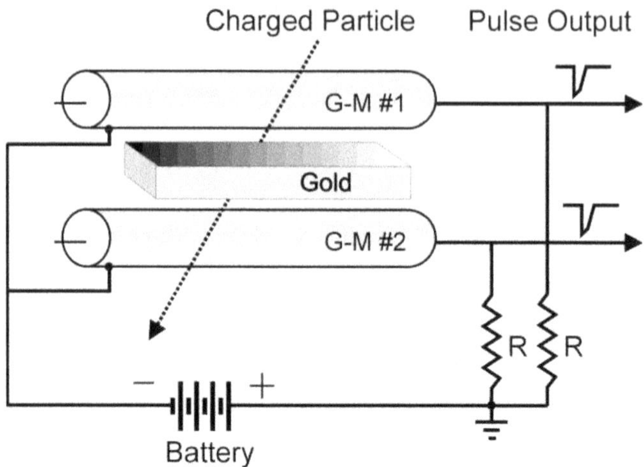

Figure 4.7 Bothe and Kolhörster gold absorber coincidence experiment.

This is an astonishing result since it was well known that only a small fraction of the assumed recoil electrons at any height in the atmosphere could have that range. Thus Bothe and Kolhörster suggested that the current ideas concerning the nature of cosmic rays were probably wrong. Moreover, they concluded that the penetrating rays consisted of ionizing particles and that the ionizing particles observed near sea level were among the primary particles that were capable of traversing the atmosphere. This was the first major challenge to the gamma ray assumption.

4.6 Conflicting Hypotheses about the Primary Radiation

Before the Bothe and Kolhörster experiments were published (Section 4.5), most physicists were of the opinion that the penetrating radiation was due to *energetic photons*. But nearly all authors studying the subject also referred to the possibility, in principle, of an electron radiation. For example, Dmitry Skobeltzyn in his fine Wilson photographs (Section 4.2), found electron tracks that were not of local radioactive origin and with energies of 15 MeV or more. One must assume that these fast electrons are somehow connected to the primary radiation. The essential question now was whether the ionizing radiation found at sea level was due to secondary recoil electrons from gamma radiation, as was assumed, or does it represent in itself the primary radiation? At this time it was impossible to distinguish between the two hypotheses. But that was about to change.

Keep in mind, that at this time (1929), high-energy photons were thought to be absorbed only in Compton collisions while charged particles were thought to lose energy only through ionization of matter.

The basis of the *photon case* rested with the sea level absorption curve that predicted a mean free path of 300 cm in water. Calculations based on the new theory of Klein and Nishina indicated energy of about 60 MeV for those photons with a mean free path of 300 cm. Subsequently, during a Compton collision of maximum interaction, the photon will disappear giving all 60 MeV of its energy to an electron. This 60 MeV electron, because it is charged, will have a mean free path of only 30 cm in water. Therefore, at the surface of the Earth there would be a mixture of photons and high-speed electrons, of which the electrons would produce the ionization effects that would be observed in an ion chamber or single G-M counter.

The basis of the *charged particle case* rested on the assumption that the primary radiation consisted of various high-energy charged particles, many of which could penetrate the entire atmosphere. That being the case, they would need to have a range of very large energies in order to make it to the surface to be detected.

Clearly, an observer at sea level would detect ionizing radiation from *either scenario*. Hence, if one measured the radiation, the results would be about the same in either case. So, although both hypotheses lead to entirely different pictures of the radiation after it has passed through the atmosphere, *there was no way of distinguishing between them.*

After the Bothe and Kolhörster experiments were published (Section 4.5), there was afforded, for the first time, the possibility of distinguishing between the two conflicting hypotheses. Recall that their experiment used two G-M counters with a block of gold 4.1 cm thick between the counters. If all the primary cosmic rays are photons of 60 MeV then no coincidences would have been found with 4.1 cm of gold between the G-M counters, since 60 MeV recoil electrons cannot travel that far in gold. On the other hand, if the primary cosmic rays are charged particles, then in order to travel as far as they did through the atmosphere they must have energies much higher than 60 MeV. If so, they could have passed through 4.1 cm of gold and caused the observed coincidences.

Therefore Bothe and Kolhörster concluded that the existing ideas concerning the character of cosmic rays were probably incorrect. In their view the penetrating rays consisted *only* of ionizing particles and consequently the ionizing particles observed near sea level were among the primary particles that were capable of traversing the atmosphere.

It is true that proponents of the photon hypothesis thought the radiation was a mixture of photons of different energies, with some even greater than 60 MeV. Hence a small fraction would have made it through the gold block. Although this makes the argument a little less exact it does not change the conclusion.

Bothe and Kolhörster realized that energies of 1000 MeV to 2000 MeV were needed to give charged particles such long range in matter and hence they might have unfamiliar properties. And this posed a serious objection to their conclusions since their interpretation was based on an arbitrary extrapolation of the known properties of photons and electrons at low energies. Conceivably, the energies of photons could also be much greater than computed from the current theory, which was limited to 1 MeV. In that case, the secondary electrons might have been able to penetrate the gold block producing the same coincidence as primary energetic electrons.

For this reason one cannot say that Bothe and Kolhörster's claim was correct in suggesting that the primary radiation consisted of charged particles. However their work was a landmark in the history of cosmic ray research. It was the first time that scientists had attempted to use laboratory experiments, not just observations, to determine the nature of cosmic rays.

As time passed, these results, and similar ones done by Bruno Rossi and others, as well as those studies done with cloud chambers were showing much more support for the charged particle theory. For example, Rossi found that, of the particles able to traverse 25 cm of lead, 60% were also able to traverse 1 meter of lead. Since all known beta rays from radioactive substances could traverse only a fraction of a millimeter in lead this was an astounding result. It meant that cosmic ray particles were very energetic and must possess energies in the range of billions of electron volts.

Also arriving on the scene was a new type of cloud chamber, the G-M counter controlled cloud chamber, first constructed in 1932 at Cambridge University by Blackett and Occhialini. It used two counters to control the expansion of a cloud chamber simultaneously with the arrival of a cosmic ray (Section 4.2).

One of the many unexpected things it discovered were cosmic ray showers, a new phenomenon, that showed cascades of secondary particles attributed to a collision of a powerful cosmic ray in matter.

In 1933 Blackett and Occhialini, were the first to observe tracks of many particles that resulted from the interaction of a single high-energy cosmic ray somewhere near the cloud chamber. A photograph of this event is shown in Figure 4.8.

The photograph was taken in a vertical cloud chamber, 13 cm. in diameter and 3 cm. deep, controlled by two Geiger counters each 10 cm. long and 2 cm. in diameter, placed above and below the chamber. It had a light aluminum piston attached to the back by a rubber diaphragm, and it was filled with oxygen to an initial pressure of 1.7 atmospheres. Water vapor was used as the condensant.

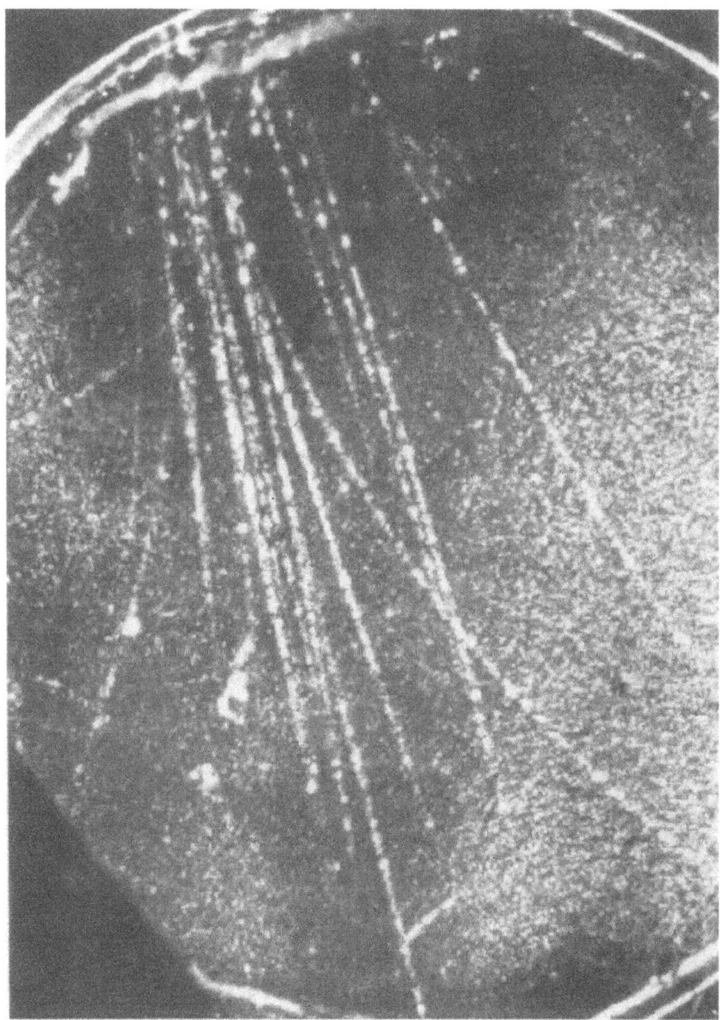

Figure 4.8 This photograph was the first obtained of an electron shower in which positive and negative electrons were identified. It was acquired in 1933 by Blackett and Occhialini with their new cloud chamber that included a magnetic field which made charged particles curve. The two most strongly curved tracks to the left correspond to negative electrons of momenta approximately 15 MeV/c, and the two most strongly curved tracks to the right to positive electrons of momenta 12 and 45 MeV/c. respectively.

The expansion took place in 0.01 sec, giving a track breadth of about 0-8 mm. in the chamber. A clearing field of 3-4 volt per cm was maintained between the top and bottom of the chamber, which was placed in a water-cooled solenoid giving a uniform field of 3,100 gauss. The direction of the field was away from the camera or in this case into the paper.

A shower of 16 positive and negative electrons diverges from a point in the copper above the chamber. The curved tracks to the left are due to negative electrons and the two most strongly curved tracks to the right of positive electrons.

It was a great disappointment for many physicists when, in the face of mounting experimental evidence such as this, that they were forced to abandon Millikan's fascinating theory of the origin of cosmic rays (Section 4.1).

But Millikan, being strong minded, tried to discredit this theory of cosmic rays and Geiger counters and wrote many papers critical of the charged particle scenario. But it was for naught, as little by little the tide was shifting; and it was away from the gamma ray assumption.

The next question to be settled was how to verify that the primary particles were indeed charged. This led to some brilliant thinking by scientists and, consequently, many new ideas surfaced including using the Earth as a magnet to determine if the primary radiation was charged. These ideas and more will be discussed in the next chapters.

Chapter 5

5 Using Earth's Magnetic Field to Study Cosmic Rays

5.1 Why Is the Earth's Magnetic Field Relevant?

As the story of cosmic rays has been unfolding so far, we have reached the stage where it seemed likely that the *primary radiation* impacting Earth consisted of highly energetic charged particles. But scientists of this time did not have access to space probes and satellites to perform experiments outside of our atmosphere as is common today. So, they obtained much of their information by using other means, such as the geomagnetic field, which was accessible on the surface of the Earth. But why was the Earth's magnetic field a factor?

Some scientists believed that if the primary cosmic rays were indeed charged, the magnetic field might influence them in some predictable way. On the other hand, if they are not charged particles, like photons, they will proceed through the field without deviation. To understand this situation more clearly, we first need to take a look at the nature of the geomagnetic field.

Swirling currents of molten iron deep in the Earth's core are responsible for this natural magnet whose influence extends more than 40,000 miles out into space on the side facing the Sun, and all the way to the Moon's orbit, at 238,000 miles on the opposite side. This *asymmetry* is caused by the *solar wind*, which is a stream of energized, charged particles, mainly electrons and protons, flowing outward from the Sun, through the solar system at speeds as high as 550 mi/sec (900 km/sec).

Figure 5.1 shows the distortion of the magnetic field lines due to the solar wind as well as a zone called the *Van Allen radiation belts*. A radiation belt is a region of energetic charged particles, most of which originate from the solar wind that is captured by and held around a planet by that planet's magnetic field. The Earth has two such belts and sometimes others may be temporarily created. The solar wind is also responsible for the aurora near the Earth's poles.

For scientists of the time a much *simpler view* was taken. For them, the Earth's magnetic field looked like that which would be produced by a bar magnet at the center of the Earth, with the *north magnetic pole* toward the south geographic pole and the *south magnetic* pole toward the north geographic pole, as shown in Figure 5.2. This picture does indeed describe the magnetic field up to a distance of about five Earth radii from the center. The magnetic dipole center is displaced 200 miles from the center of the Earth and its axis is tilted at an angle of 11.7 degrees from the rotational axis. Thus, the geomagnetic equator is skewed with respect to the geographical equator.

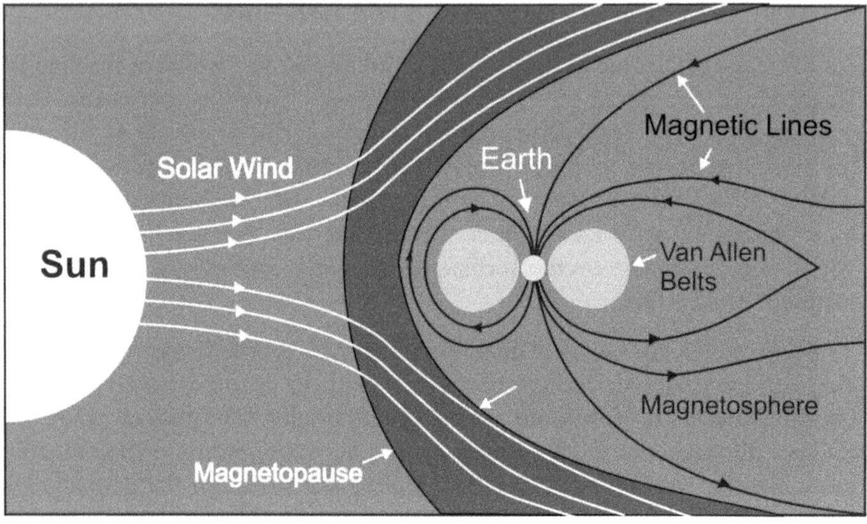

Figure 5.1 Solar wind impinging on Earth causing asymmetry in magnetic field.

Magnetic field lines loop out of the north geomagnetic pole and into the south geomagnetic pole. The lines are close together near the magnetic poles where the magnetic force is strong, and spread out where it is relatively weak. The field itself extends far beyond the atmosphere and diminishes gradually. At 1000 miles its strength is still one-half of its value on the surface. Note, by comparison, that only 1 per cent of the air mass of the atmosphere is above 20 miles.

Hence, it should be clear that if primary cosmic rays were coming towards the Earth from space they would encounter the geomagnetic field well *before colliding* with any air molecules in the atmosphere. Moreover, if the primary particles are *charged*, the geomagnetic field will deflect them, in one direction or another, depending on whether the charge is positive or negative. Neutral particles like neutrons or photons will be unaffected and pass through the field undeflected.

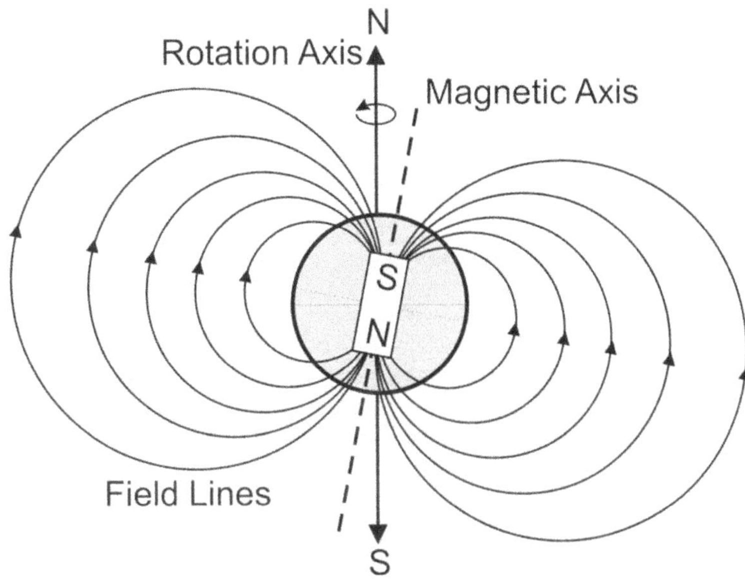

Figure 5.2 Simple model of Earth's magnetic field resembles that of a bar magnet.

The Earth's magnetic field is very small, thousands of times less than fields produced by cloud chamber magnets (Section 4.2), but its enormous extent makes up for this weakness. Thus, large deflections should be anticipated for charged cosmic rays.

To understand much of the work in the next sections we need to learn some basic facts concerning the motion of charged particles in a magnetic field.

5.2 Charged particles and magnetic fields.

As we have mentioned, a moving particle, *if it has charge*, will experience a deflecting force when moving through a magnetic field. This is called a *Lorentz force*. This force is *perpendicular* to the direction of the charge and also perpendicular to the direction of the magnetic field. It is a *vector* cross product of velocity **v** and magnetic field **B**. Because this force is always orthogonal to the particle velocity it does not change the particle's energy and only serves to deflect it in one direction or another depending on whether the charge is positive or negative.

More specifically the force is: $\mathbf{F} = q\mathbf{v} \times \mathbf{B}$, where q is the charge, **v** is the velocity and **B** is the magnetic field. We will not be solving this equation here but scientists of the time solved it for special cases. One such special case is the movement of an electron (negative charge) in a magnetic field as shown in Figure 5.3.

For this case the direction of force is toward the center of the circle while the electron orbits in a counter-clockwise circle. Notice that the force effectively *counteracts* the centrifugal force to maintain the orbit of radius R.

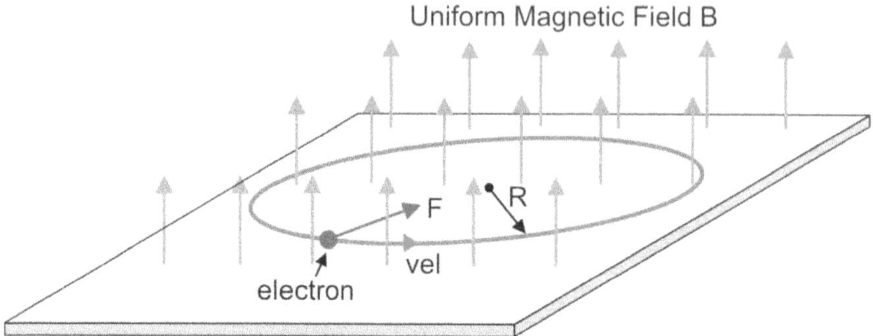

Figure 5.3 Motion of an electron in a uniform magnetic field with radius R. It experiences a Lorentz force F directed to the center of the circle.

The radius R will *decrease* with smaller velocities or higher magnetic field strengths. Because of symmetry in the Lorentz equation a similar situation arises for *positive* particles, except these particles will travel in a *clockwise* circle. Since a particle has momentum **p**=m**v**, where m is mass and **v** is the velocity of particle, we can say that the radius R depends on the momentum.

In either case, for particles of positive or negative charge, the radius of the circle R is *directly* proportional to the momentum and *inversely* proportional to the magnetic field strength B of the particle. Hence it is convenient to define the product BR, called the *magnetic rigidity*. It is a measure of the momentum of the particle, and it refers to the fact that a higher momentum particle will have a higher resistance to deflection by a magnetic field. The rigidity parameter BR is very useful in analyzing particles in the geomagnetic field.

B is often measured in *gauss* and R in centimeters so that BR is measured in gauss-cm. In looking at how energy was related to BR, scientists found a linear relationship between BR and energy E for sufficiently large values of kinetic energy. They found that E =300BR for all singly charged particles whose kinetic energy is large compared to their rest energy, where E is the energy measured in electron volts (eV).

Consider now a charged particle *entering* the geomagnetic field at the geomagnetic equator. Assume the radius R of the Earth is 6.38 x 10^8 cm and the field strength at the equator B is 0.32 gauss. This yields BR= 2 x 10^8 gauss-cm. According to the relationship for E above, a rigidity of this value corresponds to energy of 60 x 10^9 eV. Therefore, charged particles with energies of less than 60 GeV would be strongly deflected by the Earth's magnetic field.

Now, having shown that the geomagnetic field affects charged particles, let's see if we can determine the direction of deflection of the particle. Consider a *positive particle* circling the Earth above the atmosphere at the geomagnetic equator as shown in Figure 5.4. In order for the particle to be in orbit the force exerted by the magnetic field must point towards the center of the Earth.

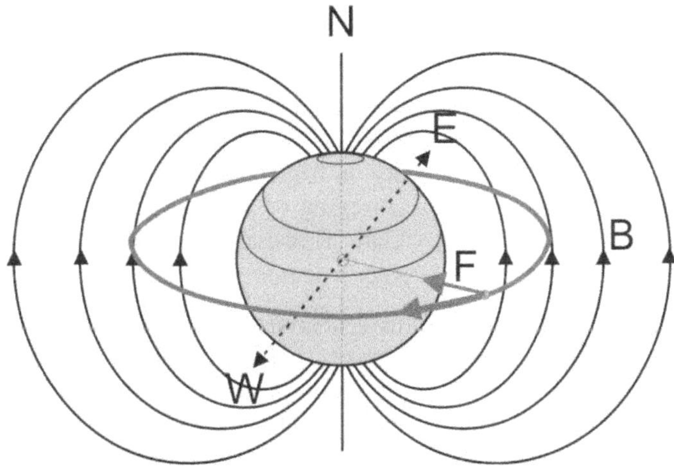

Figure 5.4 Orbit of a positive particle at the geomagnetic equator in the geomagnetic field. For this case it moves east to west.

From the previous discussion of the Lorentz force, the positive particle must be traveling in a westward direction. On the other hand, if the particle is negative it must move from west to east.

As calculated above, the magnetic rigidity for this orbit is 2 x 10^8 gauss-cm. Particles below this rigidity, entering at the geomagnetic equator, will deflect away from Earth while those above will continue towards the Earth. For each point in the magnetosphere there will be a minimum rigidity (called the cut-off or threshold rigidity) required to reach that point. Particles with less rigidity than the cut-off will be deflected away before they reach the point, whereas those with more than the cut-off may penetrate to it.

This example is a *special case* of a particle *entering* the geomagnetic field at the geomagnetic equator in a specific direction. It is used to help gain insight into the nature of the motion of a charged particle approaching Earth. The general situation is more complicated as we will see later on.

5.3 Searching For Global Cosmic Ray Variations.

Before 1927 the prevailing view of scientists was that the primary cosmic rays were photons. As we have seen in Section 4 the evidence in favor of this assumption was justified because gamma rays were more penetrating than electrons of the same energy. Hence, most scientists were not concerned about the Earth's geomagnetic field since magnetic fields do not interact with photons.

Nevertheless, a few scientists thought that if the primary rays were charged they would be moved by some means towards the magnetic poles along the magnetic lines of force. Consequently, at high latitudes the cosmic ray intensity would be higher than at the equatorial regions.

Possibly the first person to measure a variation in the cosmic ray intensity was the Dutch physicist Jacob Clay. He used an ionization chamber to record the intensity on trips between The Netherlands and Java Island in 1927. He found a lessening of the intensity of several percent in the vicinity of the Suez Canal. However the results of Clay were not generally accepted because other researchers had not found corroboration.

Millikan and Cameron in 1928 wrote that they had *not found* any significant changes in the intensity between Bolivia (19° S latitude) and Pasadena California (34° N latitude). Again in 1930 Millikan reported no change between Pasadena and Churchill, Canada (59° N latitude). That same year Bothe and Kohlhörster failed to find any variation with latitude in the North Sea between Hamburg (53° N latitude) and Spitzbergen (81° N latitude). The Swedish physicist Axel Corlin reported to have found a small latitude effect in the Baltic Sea from 50° N to 70° N latitude in Scandinavia.

Therefore the existence of any variation of cosmic ray intensity with latitude seemed doubtful. In any case the variation, if present at all, was very small and may not have been due to the Earth's geomagnetic field. Scientists observed with good reason that various conditions of the Earth's atmosphere at different locations could be responsible for significant changes in the cosmic ray intensity on the surface.

For Millikan and his supporters, the absence of any large variation in intensity was confirmation that primary cosmic rays were not charged. His insistence that primary cosmic rays were photons probably hindered his search for any variations, as he believed there was none to find! Even as late as 1931 he wrote that there was nothing to support latitude cosmic ray variations. For Millikan this lack of evidence corroborated his theory of cosmic ray formation (Section 4.1) and his "birth cry" idea that atoms were continuously being created in space.

Millikan and the others formed the wrong conclusion because of their specific voyages. Their voyages were along paths of nearly the same *geomagnetic latitude.* One would not expect large variations along such a path. A. H. Compton, who was not a follower of Millikan's birth cry idea, believed that an expanded search area might prove fruitful. In 1930 he organized a worldwide survey of cosmic ray intensities. More of Compton's work will be detailed in the sections that follow.

5.4 Propagation of Charged Particles through Earth's Magnetic Field.

It is difficult for any electrically charged particles originating from outside of the Earth's *magnetosphere* to enter inside since they tend to be deflected away via the Lorentz force (Section 5.1) similar to the way the solar wind is deflected in Figure 5.1. But the tendency to be deflected is opposed to some extent by the particles' momentum. Scientists use a quantity called *magnetic rigidity* (previously discussed in Section 5.1) to denote the ability of a particle to penetrate into the geomagnetic field. It is a derived parameter that combines as a product (BR) the effects of the magnetic field B and particle arc radius R. This is a way of expressing that a higher momentum particle will have a higher resistance to deflection by a magnetic field. The rigidity parameter is very useful since it tells us something about how particles behave in the magnetosphere.

For each point in the *magnetosphere* there is a cut-off rigidity required to reach that point. Charged particles of sufficiently high rigidity would only be deflected slightly by the geomagnetic field, but particles of lower rigidity would be deflected more and could possibly have very complex trajectories. Particles with rigidity lower than the cut-off, which are located *inside* the magnetosphere, may be trapped, and under particular climatic conditions they can create the polar *auroras*. But if they are *outside* the magnetosphere, they cannot penetrate inside and are reflected toward interstellar space.

To illustrate these concepts consider the case of three cosmic rays approaching Earth as shown in Figure 5.5. The low rigidity CR (1) is deflected away into space. Lower energy particles like CR (1) are very unpredictable and may even circle around and come back having complex trajectories. CR (2) with intermediate rigidity reaches the atmosphere along a less curved path. The high rigidity CR (3) is barely bent at all. In this case it travels along a nearly straight line right past Earth. Particles like CR (3) which is above the *geomagnetic cutoff* have no access to the atmosphere unless on a direct impact path. Because the direction of motion of particles changes continuously the paths may become highly convoluted. Their path is not simply a straight line such as astronomers get from tracing photons back to stars in the sky.

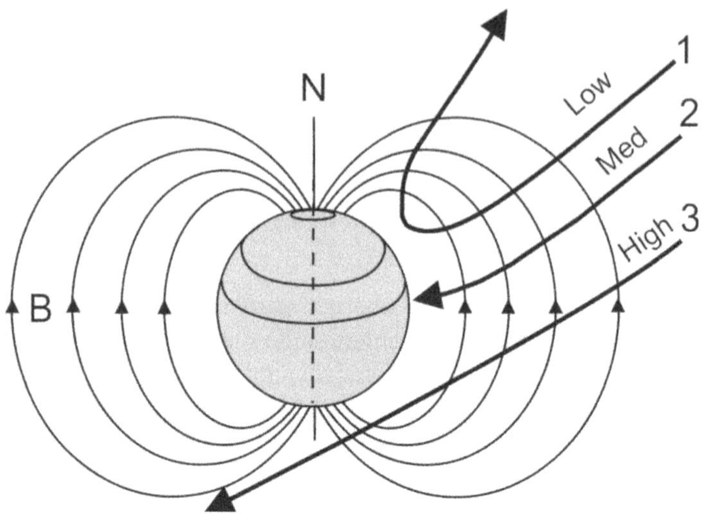

Figure 5.5 Special cases of three cosmic rays of low, medium and high magnetic rigidity approaching Earth. See text.

The precise shape of cosmic ray paths posed a great problem for scientists. There was no general formula available that could be used. But there was a theoretical background for handling the situation, based on the simplified model of the Earth's magnetic field shown earlier in Figure 5.2. This approach was first used by the Norwegian geophysicist Carl Störmer starting in 1903 when the electron was discovered. He was inspired by beautiful and striking experiments on the movement of cathode rays in magnetic fields. Cathode rays are essentially a beam of electrons emitted from the cathode inside a high-vacuum tube. He observed that they could be bent in a circle by a magnetic field.

Störmer was very interested in polar auroras, those striking luminous displays in the upper atmosphere also known as Northern Lights (Figure 5.6). The discovery of electrons and the possible association between aurora activity and solar activity fueled speculation that auroras might be generated by electrons being emitted from the sun. Thus he began his life's work of determining the classic description of electron orbits in the geomagnetic field. His research focused on low energy electrons and their trajectories.

Störmer's work produced an accurate mathematical description of how charged particles behave in a magnetic field. As happens sometimes in mathematics, there was no general analytic solution to the equations. So hc had to use painstaking numerical calculations over many years to derive a few possible trajectories. Legend has it that there were rooms full of students manually doing the computations.

Figure 5.6 Eielson Air Force Base, Alaska — The Aurora Borealis, or Northern Lights, shines above Bear Lake

Störmer's main interest was in computing trajectories of electrons with different magnetic rigidities approaching Earth from the sun. He was hoping to discover which direction these particles entered the atmosphere and which area of the Earth was impacted. By the time cosmic rays appeared on the scene, Störmer's theory was well developed. However, the application of the theory to cosmic rays was not direct because the step from the problem of the aurora to cosmic rays was not obvious at the time. The main obstacle was that cosmic rays appeared to come from all directions. A study of their paths by this method would have required the computations of vastly more trajectories which was impossible with the resources available at the time.

Many cosmic ray physicists examined Störmer's work intensely. One of them, Bruno Rossi, hit upon a different formulation of the work, which while not solving the equations, produced new insight into the problem. He conjectured that it may be easier to follow a hypothetical charged particle of given rigidity *in reverse* from the surface of Earth to outer space rather than from outer space to Earth. While this seems like an odd idea it proved profitable to consider such reverse trajectories for cosmic rays.

Consider, for example, a trajectory by a particle of certain rigidity traced backward from a point on Earth. Either it escapes to space or it returns to Earth. If it escapes from Earth it might have been a cosmic ray particle. So this would be an *allowed* direction. If, on the other hand, it returns to Earth it could not have been a cosmic ray particle. So this would be a *forbidden* direction.

Scientists of that era were not interested in trajectories as such; they wanted to know general characteristics such as which directions of arrival were allowed and which were forbidden. Fortunately Störmer's theory provided the needed answers. He had shown that there exist *bounded trajectories* that forever stayed in the vicinity of the Earth's poles. It didn't seem likely that such trajectories would be possible with cosmic rays. However, the fact that there might be even a slight chance for these bounded regions to exist for cosmic rays provided motivation for more research.

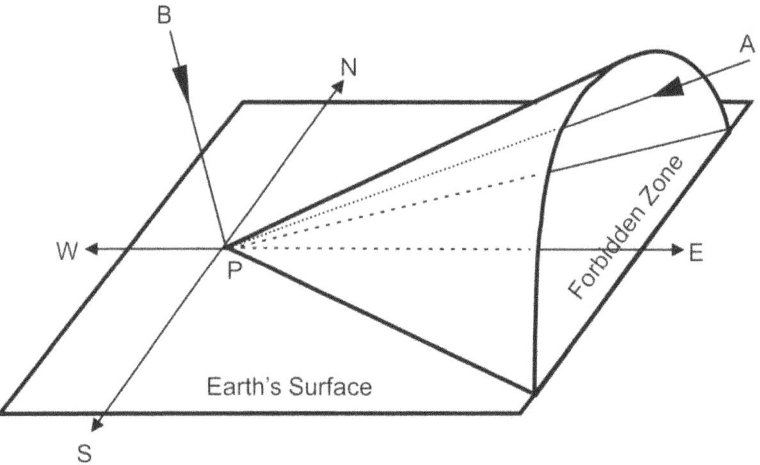

Figure 5.7 Störmer cone for positive particles. Trajectories such as AP are forbidden while those like BP are allowed. Therefore more particles would reach point P from the west than from the east.

In the course of events, Bruno Rossi using Störmer's mathematics demonstrated that for each point on the Earth, for positive particles of a given rigidity there exists a cone shaped region with its open-end facing the east such that all trajectories within the cone are bounded trajectories and thus *forbidden* as shown in Figure 5.7. Similarly, if the particle was negative, the cone would have its open-end pointed toward the west.

Although this mathematical result itself was not complete, as it left unfinished what happens to trajectories outside the cone, it was none-the-less highly useful. Specifically, it alluded to a very promising test for cosmic rays. Since the cone shaped region excluded particles of a given sign it necessarily predicted an *asymmetry* in the intensity distribution of cosmic rays reaching Earth from the sky.

For example if the Störmer cone points to the east for positive particles (thus excluding them) there should be more positive particles coming in from the west than from the east. The situation would be reversed if the Störmer cone pointed to the west for positive particles. Hence a test for extraterrestrial cosmic rays was born. It was known as the *east-west effect*.

Before we leave this important section it might be interesting to look at some more realistic trajectories of cosmic rays calculated by later researchers. In the year 2000, D. F. Smart and others used computers of the time to produce trajectories of charged particles, traced out in the vertical direction, from the same location. A typical group of *reverse* trajectories is shown in Figure 5.8. The paths undergo increased geomagnetic bending as the particle energy (rigidity) is decreased. Charged particle trajectories near the cutoff rigidity develop intermediate loops and become complex.

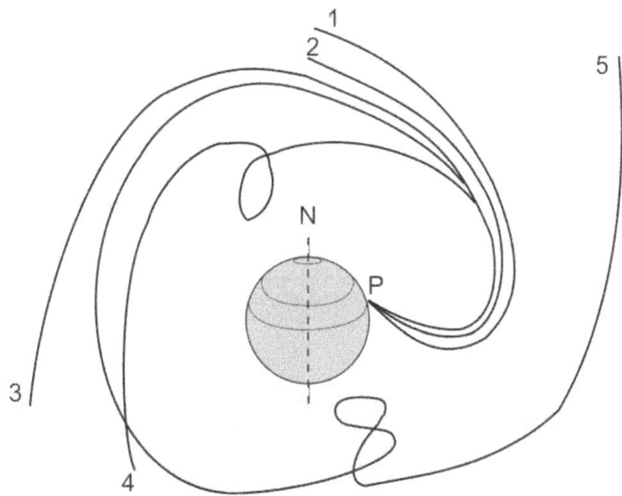

Figure 5.8 More realistic view of cosmic rays of different magnetic rigidity approaching Earth. Trajectory 1 has highest rigidity. See text.

All of this theoretical work was fine but scientists needed actual physical confirmation of the latitude effect and east-west effect. Hence around 1930 widespread experimental efforts from many scientists from many different countries were undertaken. The next section details the programs and equipment that was used in this work.

5.5 Discovery of the Latitude Variation.

As mentioned earlier, the magnetic latitude effect was first observed by Clay in 1927 in the course of a voyage from The Netherlands and Java Island where he found a cosmic ray decrease of 14 per cent. He could not offer any plausible explanation for the decrease and presumed that it was most likely due to some terrestrial cause. His finding was not accepted, mainly because it was not confirmed by other scientists (Section 5.2). In spite of this, Clay remained firmly convinced of his discovery. With more sensitive ionization chambers, he repeated his voyages several times on the motorship "Christiaan Huygens".

On his later voyages with Berlage (1932) he used an instrument constructed by E. Steinke. It consisted of a 22 liter chamber filled with carbon dioxide under 11 atmospheres pressure and surrounded by a 13 cm. shield of iron. The wall of the chamber was held at 130 volts and the needle in the ionization chamber was compensated by a known charge. He considered the accuracy to be at least 1 per cent. His readings were recorded by photographing the electrometer needle every hour.

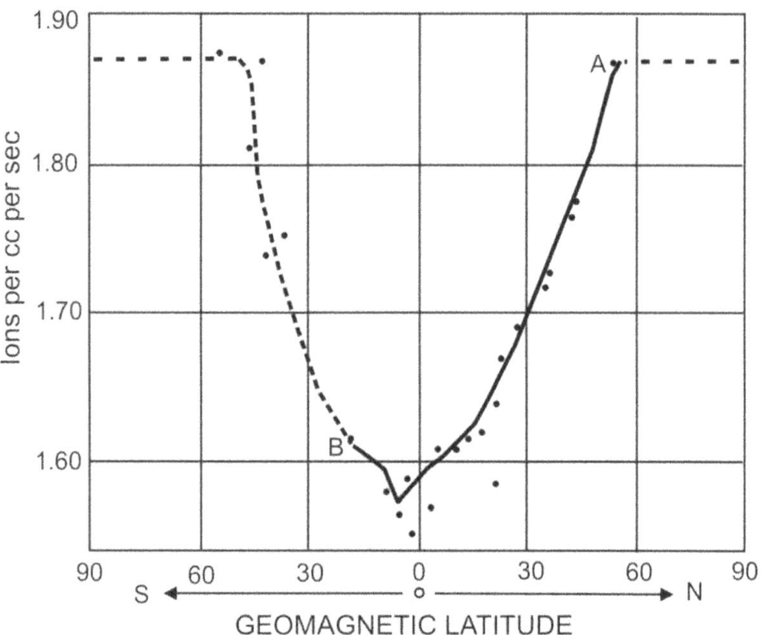

Figure 5.9 Latitude effect found by Clay in his voyages from Amsterdam (A) and Bandoeng (B). Black dots indicate where measurements were taken. See text.

Results of those measurements are shown in Figure 5.9 with the dashed lines corresponding to later readings made by Millikan, Bothe, and others for comparison. Clay obtained a reduction of 16 per cent in cosmic ray intensity from Amsterdam (A) to Bandoeng (B), which is at sea level near the geomagnetic equator. Scientific skepticism kept his work from being accepted at this time.

It was about this time in 1929 that Bothe and Kolhörster using G-M counters had shown that sea level CRs probably consisted of charged particles. They went on to point out that the Clay effect might be understood if the primary CRs also consisted of charged particles and if those particles were deflected above the atmosphere by the Earth's magnetic field. Furthermore, they were strongly convinced that direct information on the nature of CRs would be obtained by investigating the influence of the geomagnetic field on CR intensity measured at *different* geomagnetic latitudes.

Interest in the cosmic ray puzzle was beginning to attract more scientists and research was started in various parts of the world. The most intense center of it all was the University of Chicago where Arthur Compton began a "world survey of cosmic rays" (Figure 5.10). Compton was a well known person in physics having explained how high energy photons can knock electrons out of atoms (Section 3.5). This theory, known as the Compton Effect, won him the Nobel Prize in 1927.

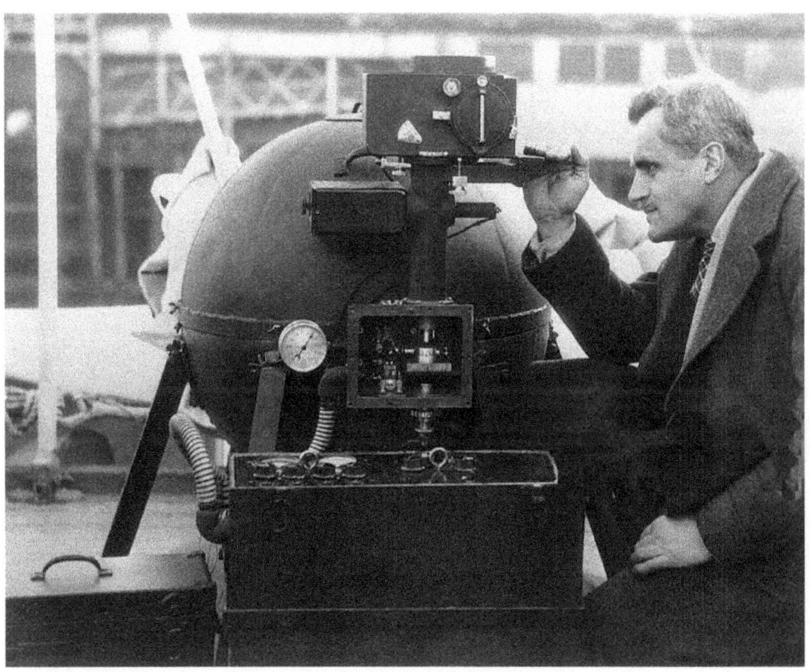

Figure 5.10 Arthur Compton on ship with his cosmic ray chamber used in the worldwide expedition survey of cosmic rays.

Preparation for this stupendous world survey was without precedent in science. Large expeditions were outfitted, the best available scientific minds were enlisted and strategy was developed on a global scale. The climax of action was reached between 1931 and 1934 when 12 expeditions from the great Chicago institution scattered to their observational points. In their travels, the crews covered 250,000 miles.

One went to South Africa, another to Peru, a third to Alaska, a fourth to Australia and so on. They visited the tropic and arctic regions alike. The Earth was used as a gigantic magnet and the mountain peaks as laboratories. Cosmic radiations were measured at high altitude and compared with those on the surface of the Earth and deep in coal mines. There were eighty cooperating physicists, making measurements at more than a hundred stations widely distributed over the Earth.

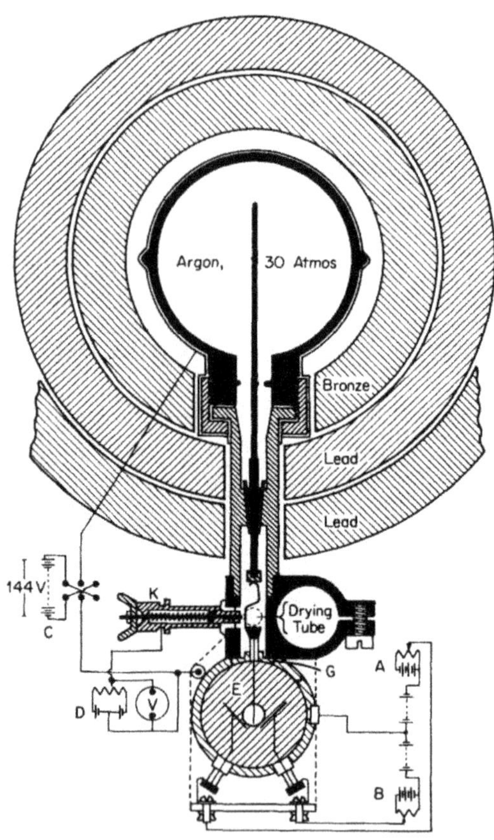

Figure 5.11 Portable cosmic ray chamber used for worldwide surveys organized by Compton in 1931. It was filled with 30 atmospheres of Argon.

They used an improved portable cosmic-ray meter designed by Compton. It had a spherical steel chamber, 10 cm in diameter, surrounded by three screens of lead and bronze. It was filled with 30 atmospheres of argon and used a standard Lindeman electrometer. Compton prepared explicit directions for its use. This made comparisons of data much easier as well as providing systematic ways of minimizing errors. A typical Compton cosmic ray ion chamber is shown in Figure 5.11.

A summary of the published observations of the various expeditions that have studied the geographic distribution of the cosmic rays at sea-level is shown in Figure 5.12. Data are shown in the form of curves (isocosms) of equal cosmic ray intensity, with the results of the various investigators all reduced to the same scale. The *solid dots* represent the sea-level locations occupied by the observers of the Chicago expeditions. The *open circles* are results reported by other observers, among them Clay, Hoerlin, Millikan and Neher, Prins, and their collaborators.

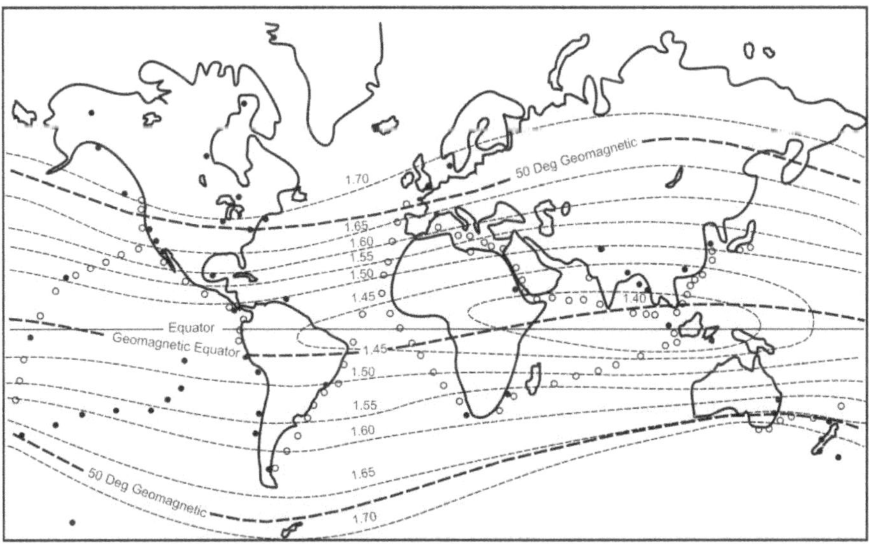

Figure 5.12 Isocosms or curves of equal cosmic ray intensity according to Compton. Numbers on the curves are intensity values. Dots show locations where measurements were made. See text.

Notice that the intensity of cosmic radiation varies not only with latitude but also *longitude*. For example the intensity along the geomagnetic equator falls to a minimum in the Indian Ocean. This occurs because the center of the Earth's magnetic dipole does not coincide with the geographic center. It is displaced toward the Indian Ocean where the magnetic field is stronger which, in turn, producces a stronger decrease in cosmic ray intensity.

Measurements among the various observers fitted the isocosms well. The curvature of these isocosms follows closely the parallels of geomagnetic latitude, identifying the phenomenon as one dependent upon the Earth's magnetic field. Compton noted that: "*Such details leave no doubt but that the latitude effect is due to the action of the earth's magnetic field. The existence of the effect, accordingly, implies that at least a considerable portion of the primary cosmic rays is electrically charged.*"

Soon after the initial data was made public, G. Lemaitre and M.Vallarta in a Physical Review article in 1933 stated that: "In the course of a survey of the intensity of cosmic radiation at a large number of stations scattered all over the world, A. H. Compton and his collaborators discovered the remarkable fact that while the intensity is nearly constant for latitudes north of 34° in the American continent and south of 34° in Australasia, it drops sharply between these latitudes to a value about 87 percent as great as that for high latitudes, reaching a minimum at or near the magnetic equator. A close correlation with magnetic latitude was also found. This discovery rules out the hypothesis that the cosmic radiation consists of photons alone and suggests that it is made up at least partly of electrons, protons or other charged particles."

R. A. Millikan, the other great name in cosmic rays at that time, whose extensive observations had convinced him that the primary cosmic radiation, incident on the Earth's outer atmosphere, was photonic in nature, was not convinced of the effect. He continued to take measurements, perhaps hoping to demonstrate the absence of the effect. As late as 1932 he was questioning the effect's existence. When mentioning Compton's earlier results he often equivocated.

As the controversy increased between the two men, Millikan sent his junior colleague Victor Neher on another latitude survey from Los Angeles to South America in November of 1932. Neher had designed a special 2000 cubic cm, self-recording anti-vibration ion chamber with an internal quartz crystal detector that would allow it to work in unusual conditions. In preparation for the trip Neher decided to take a second system in case the first one failed. But there was no time to calibrate the second system.

After leaving Los Angeles harbor all went well for a day. Then symptoms developed in the chamber and finally on the second day the quartz system broke. Neher had the second quartz system but the seas were too rough to perform the delicate calibration required until they reached Mazatlan in the Mexican state of Sinaloa. Here, in the quiet part of the bay, Neher managed to adjust the delicate balancing of the second quartz unit using a pair of cuticle shears and a magnifying glass. A day after leaving Mazatlan he began taking data again, hopeful that it was calibrated correctly. The data showed no latitude effect on the remaining trip to Panama. As arranged with Millikan, he was to send a *brief* cable from Panama. To the cost conscious Millikan, brief meant fewest possible words. Neher could not describe all the events briefly, so he decided to send a cable saying there was no latitude effect, with a letter to follow.

Hearing this news, Millikan called a press conference and triumphantly proclaimed to reporters that there was *no latitude effect*. This was followed by a historic debate with Compton at a December meeting of scientists in 1932 about the nature of cosmic rays. It received wide publicity, having gotten on the pages of the *Pasadena Star News* where an account of the debate of two Nobel Prize winners was given. The headline below their photos read, "*Cosmic Radiation Foes Battle Over Theories of Origin*".

Meanwhile, Neher proceeded northward from Mollendo and took data all the way to New York via the Panama Canal. Here it became clear that there *was* a latitude effect, as he recorded a seven per cent change in intensity, which was concealed before by a broken system and different ships. This caused some anguish for Millikan, as he now had to reconsider his previous comments.

Stubbornly, Millikan was still not completely convinced by this result. Therefore, he and Neher organized another extensive investigation on the intensity of cosmic rays on the Earth's surface at sea level. They sent instruments to many parts of the globe on 12 different voyages. These extensive voyages produced results that were substantially the same as Compton's data. When they were published in 1936, Compton, for completeness, included them with his data as shown in Figure 5.12.

The results of this period were summarized when Compton wrote an article for the *Review of Scientific Instruments*, in 1936. He decried Millikan's "birth cry" idea (Section 4.1) that the main part of cosmic rays consisted of photons, emitted when protons and electrons combine in interstellar space to form helium nuclei, and this was made part of a theory of a cyclic and eternal universe. He said that evidence has become overwhelming that most of the rays are electrically charged and of a higher order of energy than was then supposed. Therefore, he said, we need to abandon this hypothesis.

But Millikan, while now acknowledging the latitude effect, was stubbornly clinging to his "birth cry" theory. He attempted to use the existence of the latitude effect as support for his new hypothesis. He claimed that the primary rays provided bands of energy having a step behavior that worked in much the same way as his old, now discredited, gamma ray energy bands. But these attempts fell short and he was finally forced to accept the latitude effect without the "birth cry" concept.

Ultimately, evidence of the latitude effect was too strong and could not be refuted. Most scientists therefore concluded that charged particles played a dominant role in the influx from outer space. Only one more question remained. What were these particles and what was their charge? This will be explored in the next section.

5.6 Confirmation of the East-West Effect.

At about the same time as A. H. Compton was organizing a world-wide survey of cosmic ray intensities other scientists, notably Bruno Rossi, working in his laboratory in Florence, Italy was trying to take advantage of the east-west asymmetry (Section 5.3) to provide an alternate way of investigating the nature of the cosmic rays. According to Rossi there were two reasons the east-west asymmetry was a better test.

First, it did not involve a comparison of cosmic ray intensities at two different locations. This removes obvious sources of error such as variations in the atmosphere which might alter the intensity on the ground. Secondly, the sign of the charged particle would be revealed.

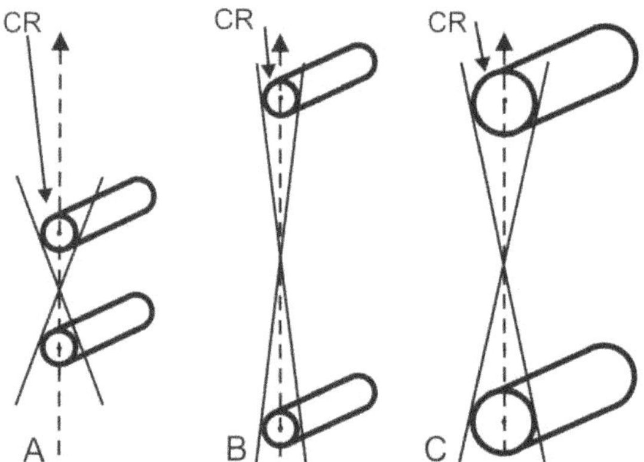

Figure 5.13 Wide angle (A) and narrow angle (B) cosmic ray telescopes. G-M tubes are connected in coincidence to indicate arrival of CR. Dashed line indicates axis of telescope. Larger diameter G-M tubes (C) increase the aperture. Only CRs within the view defined by the tubes will be counted.

Rossi, who was well versed with G-M counters, thought that they could be employed in this task. He developed an apparatus using two or more G-M tubes in coincidence with their centers aligned on an axis line. Only CRs coming from directions within the narrow aperture around the axis line formed by the tubes would produce simultaneous pulses in the G-M tubes.

This arrangement was later called a *cosmic ray telescope*. Figure 5.13 shows wide angle and narrow angle arrangements of such a telescope as well as the effect of using larger diameter tubes.

By pointing the CR telescope in different directions it was possible to compare numbers of CR particles arriving from different directions. The field of view is not circular, as an optical telescope, but depended upon the size and separation of the G-M tubes.

Rossi devised an experiment where he pointed the CR telescope alternately to the east and then to the west of the geomagnetic meridian (Figure 5.14). But Rossi was unable to detect any significant difference in the counts found in the two directions.

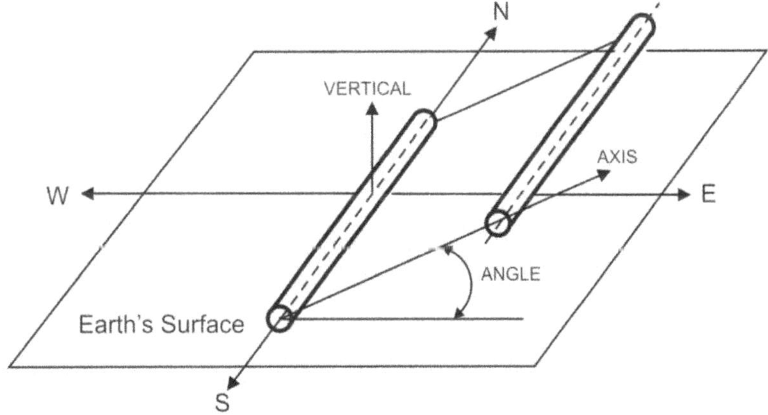

Figure 5.14 Rossi's cosmic ray telescope arrangement using G-M tubes for studying the east-west effect. It is shown here pointed to the east.

Although discouraged, Rossi remained confident in this method and planned an expedition to Eritrea, an Italian colony in Africa, where he hoped the effect would be larger. However, he suddenly became very busy at the university in Padua, so he had to postpone the trip until a later time.

In the mean time Lemaitre and Vallarta on the basis of the intensity measurements of Compton predicted that an asymmetry should be detectable in geomagnetic latitudes between the equator and 34°, with better prospects between 20 and 30 degrees. Vallarta invited both Compton from Chicago and Thomas H. Johnson at the Bartol Institute to come to the lower latitude and higher altitude of his native Mexico City where he calculated the east-west effect would be large.

Compton called upon his graduate student, Luis Alvarez, who was very familiar with G-M tubes, to construct the equipment. Alvarez worked round the clock preparing the telescopes. Johnson already had his three tube cosmic ray apparatus ready from a previous unrewarding east-west experiment on top of Mount Washington in New Hampshire.

In Mexico City, the Compton and Johnson groups setup their equipment on the roof of the Hotel Geneva at 29° N geomagnetic latitude. Results showed that when the cosmic ray telescope was inclined to 45°of the zenith, the intensity of CRs was greater from the west than from the east by a difference of 10 per cent.

Both Johnson and Alvarez and Compton published their results back-to-back in the same issue of Physical Review in 1933. Compton gave much of the credit to Alvarez for their joint work. Alvarez later recalled that they immediately wondered which positively charged particles were in the cosmic radiation.

Paul Dirac had predicted the existence of the *positron* in 1927 and Carl Anderson had discovered it in 1932 – so it was natural to propose, as many scientists did including Alvarez and Compton, that positrons were the primary particle with positive charge in the cosmic radiation. Later Johnson provided experiments and arguments for considering the proton as the primary particle.

A few months after the Mexico experiments, Rossi and Benedetti working in Asmara, Eritrea (11° N geomagnetic latitude) with telescope set to 45 degrees from the vertical found an excess of 26 per cent from the west direction. This result reinforced the idea that an east-west effect existed. Unfortunately, for Rossi, his result came too late to recognize him at the discoverer of it.

None-the-less the asymmetry confirmed the particle nature of the primary radiation and made it clear that a portion, possibly all, of the primary radiation consisted of positively charged particles. This was a most unexpected result since it was believed by proponents of the particle hypothesis that the primary cosmic radiation consisted of negative particles, like electrons. Up to this time the only known positive particle was the proton. But as noted above, in the same year another positive particle was discovered, the positive electron or positron. This left open the question of the true nature of the cosmic rays and their formation.

5.7 Role of the Atmosphere in Studying These Effects

From the experiments described in the last section, it was concluded that the primary rays instrumental in producing effects at sea level were largely positively charged. But the kinds of particles involved were not known nor where they came from. Scientists speculated about these questions and undertook studies to gain more knowledge about them. But, seemingly, the more they investigated, the more puzzles arose. One question concerned the role the atmosphere played in observing these effects.

Investigators were largely confined to measuring cosmic rays at or near the Earth's surface and, as might be expected, there was great interest in determining how the particles propagated through the atmosphere. The simplest picture one could imagine for explaining the measurements at sea level would be to consider a single positive charged particle passing through the atmosphere without scattering or secondary emission other than that of ionization. Recall from Section 3.3, that a charged particle may slow down by ionization but would not necessarily disappear. Thus, it would be detectable in experimenter's ionization chambers or G-M counters by virtue of the ionization it produced. In this idealized scenario, the number entering the top of the atmosphere would be the same as measured at sea level.

W. F. G. Swan considered this simple picture and concluded it would not work based on his understanding of the experimental data. He therefore proposed in 1934, a new hypothesis to the effect that the primary rays emitted long range *secondary* particles, possibly through intermediary processes. He expected the long range secondary particles to follow the primary with little deviation.

Other scientists, too, had good reasons to believe that the local radiation observed in the atmosphere consisted largely of secondary particles. But if that were true, how could observations on the Earth's surface justify a theory that supposedly applied to the primary radiation arriving at the top of the atmosphere from outer space?

For the latitude effect the answer was obvious. Since the number of secondary particles in the atmosphere would increase or decrease as the number of primary rays entering from above, secondary rays would exhibit a latitude effect very much like the primary rays. The effect would not be as pronounced since the primary particles with the largest latitude variation are those of lowest energy and likely to produce fewer secondary particles.

In this connection it is enlightening to consider the results of high altitude measurements made on mountains and by means of balloons in the early 1930s by groups associated with Compton and Millikan. A composite of results is shown in Figure 5.15. The intensity at the equator clearly varies with altitude. Its value is 1.6 ion pairs per cc per sec at sea level increasing to 5.1 at 4360 meters. Obviously, the intensity is not constant with altitude as would be expected from the simple charged particle model discussed above.

In all three curves the latitude effect is quite apparent, i.e., the intensity gradually increases starting from the equator going to the higher latitudes, and then levels off. Surprisingly, the latitude effect gets stronger as the altitude is increased. This can be explained by noting that the measuring devices are operating under a smaller atmospheric thickness (higher altitude). Hence they could detect particles of lower energy which were more influenced by the Earth's magnetic field.

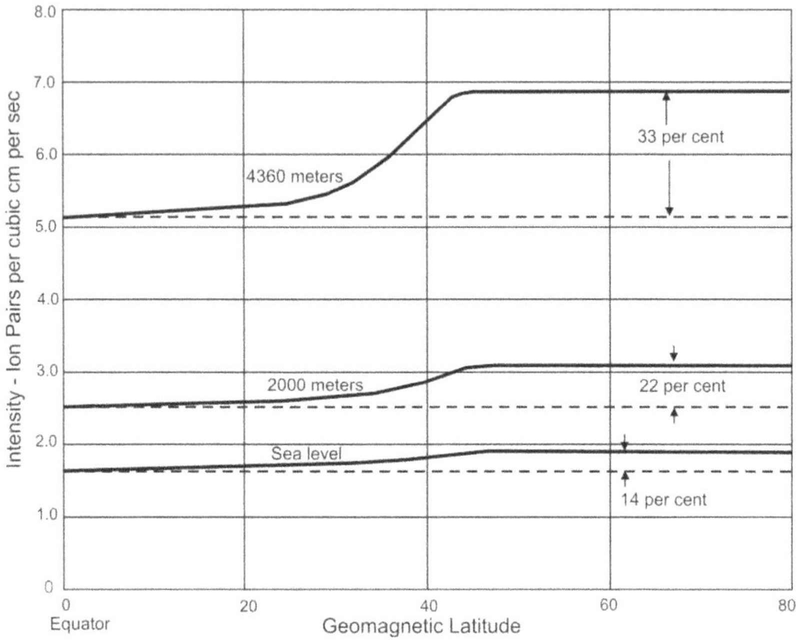

Figure 5.15 Cosmic ray intensity versus geomagnetic latitude for three different altitudes starting at sea level.

Explanation of the east-west effect requires additional considerations. Assume a primary cosmic ray colliding with the atmosphere and producing secondary particles. If these secondary particles went off at wide angles to the primary direction, the secondary radiation would be distributed, more or less uniformly in all directions regardless of the original direction of the primary. For this situation, the only hope of seeing an east-west effect would be to send the measuring instruments beyond the atmosphere where they could see the primary radiation undisturbed. But a strong east-west effect *was* found at sea level, where presumably most of the observed rays were secondary. Therefore it must be concluded that the direction of the secondary rays produced in the atmosphere must be strongly aligned with the direction of the primary rays.

Additional support for this inference comes from results observed using a cosmic ray telescope. Using this device, it had been found that the radiation that reaches us at sea-level depends upon the zenith angle θ. The zenith is the direction perpendicular to the Earth's surface. In this picture the angle θ measures the direction of the line of flight of an entering cosmic ray particle and the zenith. Such measurements are typically done in the east-west plane.

A representative plot of the zenith angle dependence is shown in Figure 5.16. As one might expect, the intensity of the radiation is maximum at zenith θ = 0 ostensibly because the steeper rays correspond to traversals of greater distances in the Earth's atmosphere. Clearly these steeper-going particles are subjected to a greater chance of absorption than the particles that are produced directly overhead. Such an explanation would hold true only if the secondary particles travelled in more or less the same direction as the primary rays that produced them. This idealized plot does not show an east-west effect.

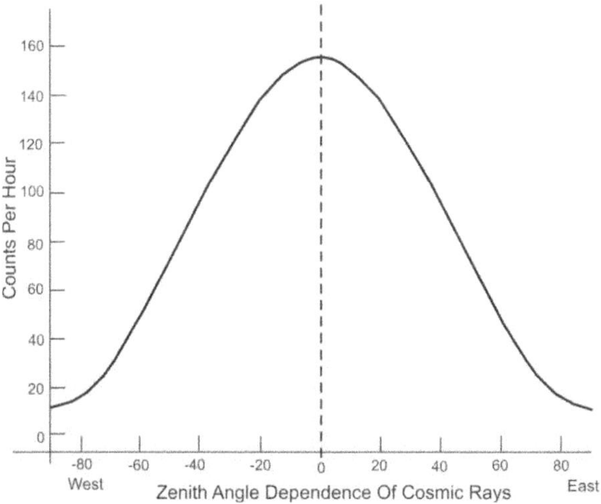

Figure 5.16 Typical cosmic ray telescope count variation with zenith angle at sea level. This idealized plot is symmetric about the zenith and the slight east-west effect is not shown.

By discovering the latitude and east-west effects scientists had begun unlocking the mysteries of cosmic rays and the structure of matter. Compton in 1936 identified two objectives for the future of cosmic ray research.

First, investigate the properties of the rays. The most immediate concern is their composition - the nature of the particles of which the rays are composed, and the energies which these particles possess. What effects do the rays produce as they traverse matter? Where do they originate and how are they produced?

Secondly, use cosmic rays as a tool in other investigations. What kinds of particles exist in nature, and how do particles of very high energy behave? Cosmic rays may also be a powerful tool in astronomy, biology and for studying the history of the universe.

Indeed, a veritable Pandora's Box had been opened by this research. As a result, an explosion of discoveries began in the late 1930's. In the next sections we will explore many of these results from this fascinating period in cosmic ray history.

Chapter 6

6 Cosmic Rays and the Emergence of Elementary Particles

6.1 The Birth of Elementary Particle Physics

To appreciate the impact of cosmic rays on the study of elementary particle physics, let's take a look back at physics in the late 1890's and see how it went from there. Then, there was only a simple view of the structure of matter. Most scientists had accepted the *atomic* idea, the notion that matter consisted of atoms. At this time, atoms were thought to be *indivisible* with nothing lighter than a hydrogen atom expected to exist. (Also see Section 3.1). But this was about to change radically with a flurry of scientific discoveries often referred to as the origin of nuclear physics. This ultimately led to the birth of the field of particle physics.

It began with the accidental discovery of *strange* rays in 1895 by Wilhelm Conrad Röntgen of Germany while conducting experiments with cathode rays. Like many scientists he was on the lookout for new phenomenon. This involved working with partially evacuated tubes, similar to fluorescent lights, called Crookes tubes, filled with a special gas (Figure 6.1). In this case he was repeating an experiment that others had performed using a Lenard tube, a variation of the Crookes tube with a thin aluminum window that permitted cathode rays to exit the tube.

His experiments involved passing electric current through the gas using a Ruhmkorff induction coil to generate a high voltage. When he did this, the tube would produce a fluorescent glow. In the course of one experiment in a dark room, with the tube shielded with heavy black paper to protect the aluminum window, he could see green colored light a short distance away on a screen painted with barium platinocyanide, a special fluorescent material. Other experiments with conventional Crookes tubes covered with heavy cardboard produced similar results.

He realized that a new type of invisible light or ray was being produced from the tube, a ray that was capable of passing through the heavy paper covering it. Additional experiments showed that the new ray would pass through most substances, casting shadows of solid objects on pieces of film. He named the new ray, *x-ray*, because "x" is often used to indicate an unknown.

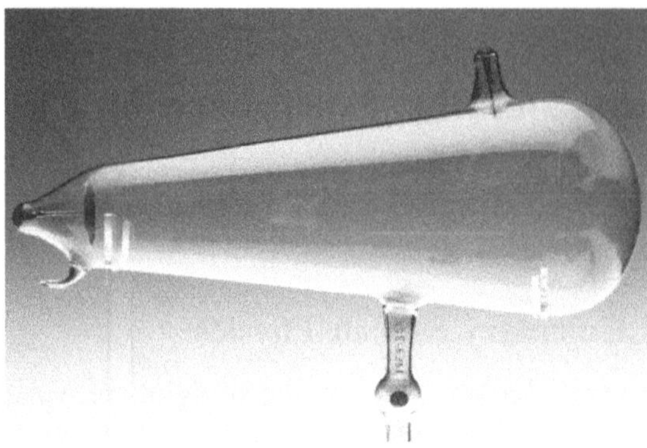

Figure 6.1 Crookes tube similar to one used by Röntgen. It is an early experimental electrical discharge tube, with partial vacuum.

In early 1896, in the wave of excitement following the discovery of x-rays, Antoine Henri Becquerel thought that phosphorescent materials, such as uranium salts, might emit penetrating x-ray-like radiation when illuminated by bright sunlight. He exposed potassium uranyl sulfate to sunlight believing the uranium might absorb the sun's energy and then re-radiate it as x-rays. Normally, he would then place the material on photographic plates to look for x-ray emissions. But, this particular day, his experiment could not be performed, because the sky was overcast. He left his unfinished experiment in a drawer over night. Since the sun did not show itself for several days, Becquerel decided to develop his photographic plates anyway. To his surprise, the images were strong and clear, proving that the uranium emitted radiation by itself. Becquerel had just discovered *radioactivity*.

So, the path that resulted in the discovery of radioactivity began with the mystery of cathode rays. While cathode rays were valuable in uncovering a new type of radiation, little was known about the cathode rays themselves. Scientists thought they might be either waves or particles. Experiments were needed to solve this quandary.

In 1897 Joseph J.Thomson was studying electric discharges at the well-known Cavendish laboratory in Cambridge, England. It was known that the discharge and the glow in the gas were due to something coming from the cathode, the negative pole of the applied high voltage. Thomson made a series of experiments observing that the cathode rays were deflected by both electric and magnetic fields; they were obviously electrically charged. By carefully measuring how the cathode rays were deflected by the fields, Thomson came to the remarkable conclusion that the rays are negatively charged particles with a mass approximately two thousand times smaller than the mass of a hydrogen atom, the lightest atom.

Thomson consistently obtained the same result, using different materials for the cathode and gas, thereby showing there was no material dependency. As a result he concluded that the cathode rays were a *new state of matter*, a state in which the subdivision of matter is carried very much further than the atom. He had discovered the *first particle*. They are now called *electrons*, although he originally called them corpuscles. It is electrons that carry the electric current across a cathode-ray tube and give rise to the strange glow at the opposite end of the tube.

While this discovery was significant it raised a critical question as to the nature of the atom. Since the atom appears to be neutral overall, positive charges must also exist to balance the negative electron. But where are these charges and how are they distributed?

In 1904 Thomson proposed that atoms could be described as negative particles floating within a soup of diffuse positive charge, the so-called *plum pudding* model, named after a popular English dessert (Figure 6.2). Though invalid by modern standards, this model represents an important step in the development of atomic theory. Not only did it incorporate new discoveries, such as the existence of the electron, it also introduced the notion of the atom as a non-inert, divisible mass.

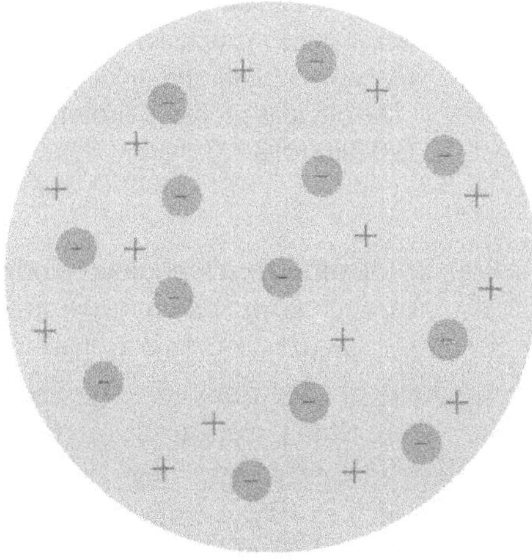

Figure 6.2 Thomson's plum pudding model of the atom.

In 1909, Ernest Rutherford at his laboratory at the University of Manchester, assigned to Ernest Marsden, a young student of Hans Geiger, the task of measuring how an alpha particle beam is scattered when it strikes a thin metal foil. Rutherford was familiar with alpha particles since he had identified and named them a decade earlier. Marsden used gold leaf and a scintillating screen to detect the scattered alphas from a radioactive source. He placed the screen not behind the gold foil, but to the side, next to the source. In this way he could detect alphas reflected back through large angles.

Marsden observed that occasionally alpha particles, instead of going straight or nearly straight were deflected at considerable angles. When Marsden related this fact to Rutherford, the Professor made him repeat the experiment to confirm it. The big *deflections* greatly amazed Rutherford. He later said that it was as if someone had fired a pistol at a piece of paper and the bullet bounced back.

Rutherford thought more deeply about these alpha deflections and in 1911 announced that he understood the findings. Moreover, he had a new explanation of atomic structure (Figure 6.3). He realized that the approaching alpha particles, with a double positive charge, must be repelled by the positive charge within the gold atoms and be deflected away. The closer the alphas approach the atom, the more they are deflected, until in extreme cases they come to a halt and are turned round in their tracks. When Rutherford calculated how close the alphas should get in this case, he found it was a tiny fraction of the atom's radius, just one ten thousandth of the atom's radius. This means that the positive charge is found only at the very center of the atom, not distributed throughout the atom as in the plum pudding model. Rutherford had discovered that atoms consist of a compact, positively charged nucleus, with the negatively charged electrons circulating at a relatively large distance.

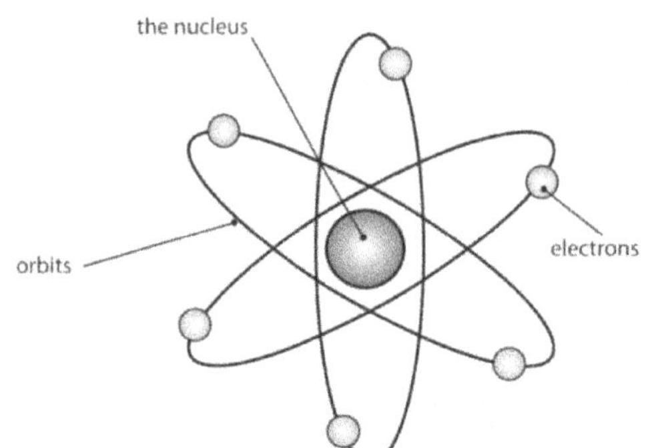

Figure 6.3 Rutherford's Nuclear Model in 1911. The model described an atom with a tiny dense center (nucleus), and electrons circulating in an orbit. The nucleus included protons and neutrons.

Rutherford's nuclear model of the atom was a huge step forward in understanding nature at the ultra small size. But it had its problems too. Since the nucleus and its cloud of electrons are oppositely charged, and therefore attract one another, there didn't seem to be anything to stop the electrons from being pulled instantly into the nucleus. Rutherford countered by saying that the atom was like a miniature solar system; the electrons circled the nucleus in wide orbits just as planets orbit the sun. The major flaw in Rutherford's model is that it describes charges that are *accelerating*. It is well known that accelerating charges *radiate* energy. In this case we have charged electrons changing direction (i.e. accelerating) as they move around their orbits. So there was nothing to stop the orbiting electrons from losing all their energy and spiraling into the nucleus. Niels Bohr resolved this problem later on with a new model of the atom.

The discovery that the positive charge of an atom is concentrated in a central nucleus raised the question of what precisely carries this charge. Rutherford's group carried on numerous experiments on many elements and discovered what they called the *H particle* common to them all. He concluded that this particle, the hydrogen nuclei, was a fundamental building brick to all elements. Later on he named it the *proton*.

Throughout the 1920s, it was the consensus of scientists that all matter was made of two elementary particles, negative electrons and positive protons. In order to maintain atom neutrality, it was further assumed that the atomic nucleus was composed of protons and "nuclear electrons". Hence a neutral atom of mass number A and charge number Z would contain in its nucleus A protons and (A – Z) electrons. There would be Z electrons in its electron cloud. All forces in the atom were believed to be electromagnetic.

For example, under this hypothesis, the nucleus of nitrogen of mass 14 would be composed of 14 protons and 7 electrons, for a net charge of +7 and a mass of 14 atomic mass units. Orbiting the nucleus would be another 7 electrons, termed external electrons.

The details are too technical to get into at this point, but this model presented several experimental and theoretical contradictions. Physicists knew of the difficulties with this view but there was not a consensus on how to revise this picture without creating more problems. Three major fields, nuclear physics, cosmic rays and quantum field theory came together to study this difficulty. By the 1930's there was conflict and paradox where these fields overlapped. It is noteworthy too, that in this seminal period, relativity and quantum mechanics also were established.

Scientists considered radical suggestions for modifying the mechanics, the electrodynamics and even the basic conservation laws in order to resolve these difficulties. But the resolution came not from these suggestions but from the introduction of new particles. These particles were the neutrino postulated by Wolfgang Pauli in 1930 and the neutron, discovered by James Chadwick in 1932. These particles allowed the elimination of electrons from the nuclei of atomic models. The new assumption is that the nucleus is composed of neutrons and protons. This nuclear model is now universally accepted. However, to reach a complete understanding of this model, it was now important to study interactions of these particles to derive more nuclear properties. This had vast consequences as we will see later on.

To summarize this section, at the start of the twentieth century there was the elegantly simple world of physics that consisted of only two particles carrying mass, one positively charged (proton) and the other negatively charged (electron). But observations were challenging this ideal, pointing to far more complexity in nature. The effort to revise this picture led to new problems as well as the birth of the field of particle physics.

6.2 Discovery of the Positron and the Creation of Matter

Cosmic ray scientists were concerned at this time with discovering the nature of cosmic rays and the special particles involved. But life was hard because little was known about either one. Some of the best minds went to work on this intricate scientific task, using skillful experimentation and imaginative theoretical thinking. These efforts were to bring solutions to the cosmic ray puzzle and produce dramatic advances in many fields of science.

The first task at hand was to determine the nature of the local cosmic radiation. As mentioned earlier (Section 4.5), Bothe and Kolhörster were the first to observe that the coincidence rate had dropped by 24 per cent when a 4.1 cm block of gold was placed between the counters. Later on, Rossi and others had shown that after passing through some thicknesses of lead that the cosmic rays became "harder", thus having the ability to traverse lead thicknesses on the order of meters.

Hence it became clear that the sea level cosmic radiation consisted of two groups (Figure 6.4). One was a soft group responsible for the steep initial drop in the absorption curve and a hard "penetrating" group corresponding to the straight part of the curve.

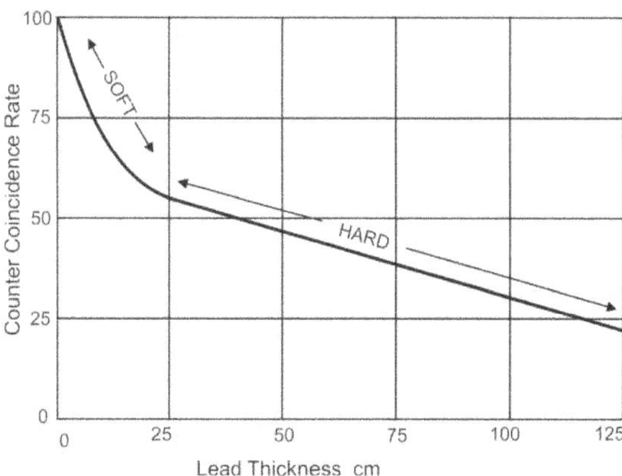

Figure 6.4 Soft and hard cosmic ray groups at sea level.

In addition to these findings a new phenomenon, *cosmic ray showers*, was discovered using cloud chambers (See Section 4.5). They were thought to be the result of the disintegration of atomic nuclei but there was no proof of this. As for the soft and hard secondary particles, they were completely mysterious, at this time.

The focus of research at the end of the 1920's changed significantly when scientists began using two new methods for studying cosmic rays and the behavior of charged particles produced by collisions of primary CRs with the atmosphere. These were the G-M coincidence counting method and the cloud chamber augmented with a magnetic field. See Sections 3.4 and 4.3 to review these concepts.

One of the major innovations was the automation of cloud chambers whereby the expansion was triggered by the passage of a cosmic ray. This *counter-controlled* chamber would activate a camera from a pulse from the cosmic ray so that the incident particle would take its own picture. This greatly improved the efficiency and reduced waste, since before this process sometimes hundreds of random photos had to be taken in order to capture a single event.

Such a lucky, random event happened in 1928 when Dmitry Skobeltzyn, working with gamma rays from radium, accidentally detected a cosmic ray in his cloud chamber (Section 4.2). His chamber was setup to observe recoil beta particle trajectories in a magnetic field. A charged particle's track, in this field, is curved with radius of curvature directly proportional to the particles momentum and inversely proportional to the field strength. Intrigued with the potential of this method, other investigators began using it to study cosmic rays.

At the California Institute of Technology in 1929 was a young researcher, Carl D. Anderson who was using a cloud chamber to study photoelectrons produced by x-rays. Upon his graduation, he wanted to stay on to study Compton collisions of nuclear gamma rays. Before he could make definite plans, he received a call from Robert Millikan, the distinguished Caltech professor, asking him to stay at Caltech and build an instrument to measure the energies of the electrons present in the cosmic radiation.

As Anderson recalled later, "it was to consist of a cloud chamber operated in a magnetic field. This machine would require a very powerful magnetic field, for the cosmic ray electrons were expected to have energies in the range of several hundred million electron volts rather than 1MeV from known radioactive samples" (Figure 6.5).

The existence, at the time, of a 450 KW DC generator, used by the Caltech aeronautical laboratory for wind tunnel tests, helped dictate the design of the new magnet. The new magnet was built using a pair of air-core solenoids capable of operating at the full power of the generator. When operated in this manner it produced a magnetic field of slightly over 24,000 Gauss. At this field strength, he was able to measure the magnetic deflection of tracks as large as 7 meters with corresponding magnetic rigidity of 1.68 x 10^7 gauss-cm. This corresponds to a kinetic energy of 5 GeV for the mass of an electron and 4 GeV for the mass of a proton – several hundred times greater than the energy limits of previous machines.

Figure 6.5 Carl Anderson and his giant cloud chamber at Caltech.

The first results from the cloud chamber were dramatic and completely unexpected. It soon became apparent that particles of the local radiation had an energy distribution ranging from low energies of 100 MeV, to above 1 GeV. In addition, there were approximately equal numbers of particles of positive and negative charges.

Initially, Anderson thought the negative particles were electrons and the positive ones were protons, since in all cases the particles carried a single charge. As more data was accumulated he was compelled to alter his position because the specific ionization (see Section 3.4) of these positive particles was too low to be interpreted as protons. His alternative interpretation was that these particles were either electrons moving upward or some unknown lightweight particle of positive charge moving downward.

To resolve this apparent paradox, a lead plate was inserted across the center of the chamber dividing it into two sections. A particle traversing the plate would lose energy and would emerge with reduced rigidity, hence disclosing its direction of travel.

According to Anderson, "It was not long after the insertion of the plate that a fine example was obtained in which a low energy lightweight particle of positive charge was observed to traverse the plate, entering the chamber from below and moving upward through the lead plate. "

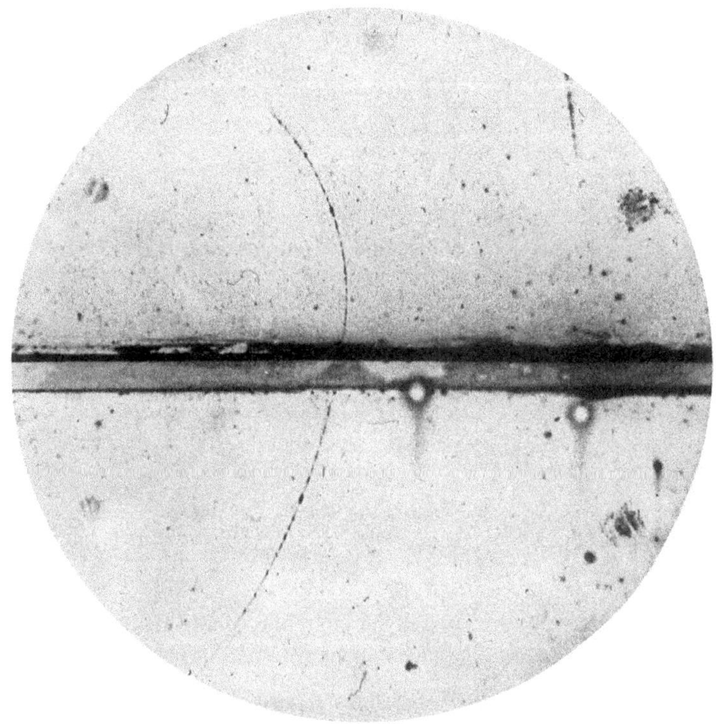

Figure 6.6 Anderson's cloud chamber positron picture. There is a lead plate midway in the chamber. The entrant particle in this case is coming in from the bottom and bends to the left.

Ionization measurements clearly showed this particle to have a mass much smaller than that of a proton and consistent with an electron mass. Curiously enough it was an example of a very rare upward moving cosmic ray particle. Without the plate, the particle would have been mistaken for an ordinary negative electron. As seen in Figure 6.6, the direction of motion of the particle is clear from the track above the plate which has a smaller radius of curvature than the part below the plate. This particle became known as a positive electron or *positron*.

Several months later, Patrick Blackett and Giuseppe Occhialini, at the Cavendish Laboratory in England, carried out some important studies using a counter-controlled cloud chamber. These experiments *confirmed* Anderson's discovery and produced other results of crucial importance.

First, they found that all shower particles deflected in their relatively weak magnetic field of 2000 gauss, were either negative or positive electrons. And for most showers, the numbers of positive and negative particles were close to the same.

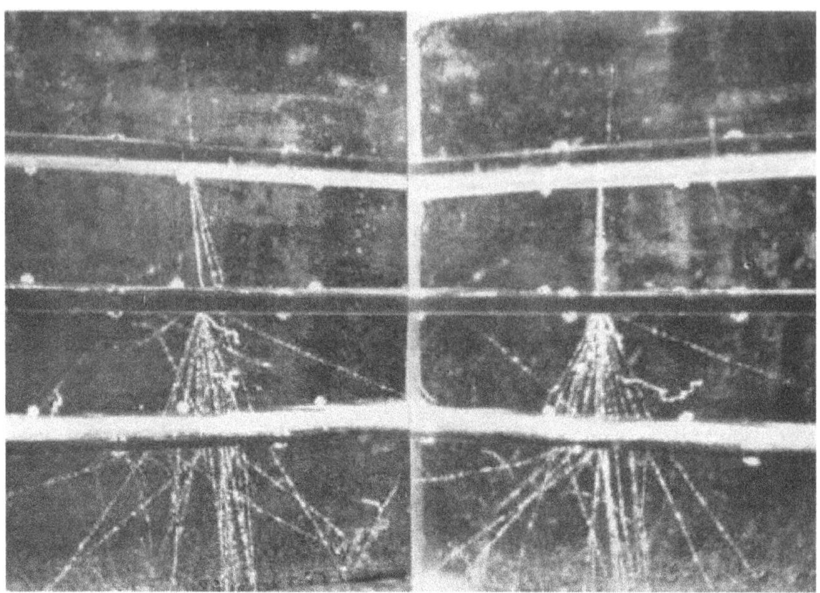

Figure 6.7 An early cloud chamber stereographic photo of an electron induced shower. An electron hits the top plate starting the small cascade. It multiplies at the two lower plates.

The two researchers took many pictures with the cloud chamber partitioned by a lead plate. Surprisingly, the photos often showed showers originating from the plate and chamber walls *simultaneously*. It seemed too, that particles generated in the showers were more likely to produce secondary showers than the ordinary cosmic ray particle. Figure 6.7 is typical of a small cascade initiated by an entrant electron in the top plate. Three or more particles leave the plate. At the second plate considerably multiplication takes place.

And finally, they also found showers that seemed to originate in the lead plate but with *no ionizing track* entering the plate from above. Therefore, such showers could not have been produced by an electron or other charged particle. In view of this evidence, it became necessary to assume the existence of *neutral* particles capable of making the showers.

At about this time in 1928, a young theorist at Cambridge University,

Paul Dirac (Figure 6.8), was working to combine the basic principles of quantum mechanics with the requirements of Einstein's theory of special relativity for electrons in electromagnetic fields. This work was later to be known as quantum electrodynamics (QED). He was well known, having developed a new theory of the electron several years earlier. His latest theory worked well in some situations, such as predicting the spin and magnetic moment of the electron, two items that were missing from non-relativistic theories. But his theory also had *big difficulties* as it predicted absurd things like particles with negative mass and negative energy.

Figure 6.8 Paul Dirac originated quantum electrodynamics (QED). QED is essentially a combination of quantum theory and special relativity. It describes how light and matter interacts.

Several months later, Dirac showed that his theory made sense if *both* positive and negative electrons existed. He could not assume the positive particle was the proton because his mathematics required the mass of the two particles to be the same. In other words each particle should be the *antiparticle* of the other. Immediately, Blackett and Occhialini recognized the positive electron in their experiments as the one required by Dirac's theory. So, at least one part of the puzzle had been solved.

It was still a mystery why positrons were so rare and eluded previous detection. Dirac's theory provided the answer. It predicted that a positron in matter had a very short life because if it met an electron, the two particles would annihilate each other. The predicted lifetime, 10^{-9} seconds, before annihilation gave a good reason why positrons had not been observed before. Their mass would be converted to energy in accordance with Einstein's equation, $E = mc^2$, with this energy being radiated as a photon.

Now, why did showers contain equal numbers of electrons and positrons? This was answered again by the symmetry of Dirac's mathematics which required that a positive electron would be produced for every negative electron produced. And, amazingly, the created mass of the two particles results from an extraordinary *materialization* process, known as *pair production*, in which all or part of the energy of the primary shower particle is transformed into mass. (Also see Section 3.5)

The non-ionizing links found by Blackett and Occhialini still needed clarification. They could be explained if the pair production process was caused by a high energy photon. Being non-ionizing, photons would not show tracks in the cloud chamber but the results of the interaction would be seen. Dirac's theory fully predicted that high energy photons could cause such a materialization of energy. More pieces of the puzzle were now being explained.

Soon after this, experiments with high energy gamma rays from radioactive sources showed that an electron and positron could be created in the field of a nucleus in such a way that the energy required to create the mass of the pair, $2\ mc^2$, along with their kinetic energy was supplied by the incident radiation. This gave additional support for the pair production hypothesis.

Lastly, scientists could now explain why high energy gamma rays were absorbed more rapidly in matter than the current theory predicted. The excess absorption can be explained by pair-production. At sufficiently high energy, photons were absorbed not only in Compton collisions but also by the materialization of photon energy into positive and negative electrons.

To summarize this period, there were many important observations of events that led to the discovery of the positron. These results were a direct experimental verification of Dirac's theory in which energy is transformed into matter and vice-versa. It stands as one of the most sterling accomplishments of science in the twentieth century.

6.3 Particle Showers and Interactions

Many of the outstanding achievements associated with the discovery of the positron, described in the previous section, still left many unanswered questions about cosmic ray showers. Dmitry Skobeltzyn first observed showers in 1929 while he was investigating beta rays from radioactive sources using a cloud chamber in a magnetic field of 1,500 gauss. On several of the photographs he noticed some tracks of high energy particles in small groups of 2, 3, or 4 tracks apparently diverging from a point somewhere near the chamber. This phenomenon immediately attracted the attention of other physicists. Skobeltzyn's work was followed up by Kunze in Kiel and by Anderson in Pasadena in 1931.

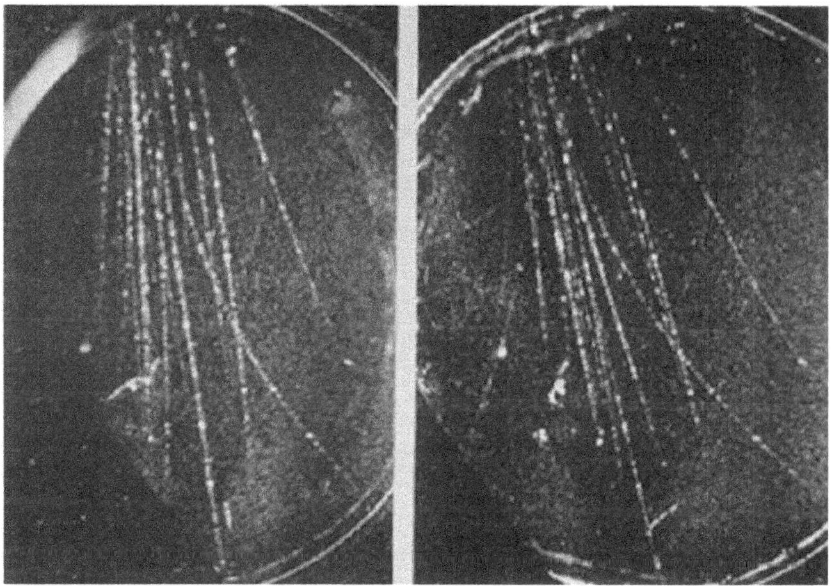

Figure 6.9 Stereographic photograph of cosmic ray shower obtained by Blackett and Occhialini in their counter controlled cloud chamber.

Later, in 1933, Blackett and Occhialini, while searching for cosmic rays, observed showers of up to 20 tracks in photographs taken with a counter-controlled cloud chamber (see Figure 6.9). They described it as a group of particles seemingly starting from one point. The intriguing part of showers is that physicists had not encountered anything remotely like it in any of their previous work.

It seemed that the showers were often the result of two or more successive collisions. But how was that possible? Indeed, this discovery led to many more questions such as, what was the nature of the interaction, did it involve atomic nuclei, and what kind of strange particles were capable of starting these showers? As we will see, new experimental studies along with new theoretical work helped provide answers to these questions.

Many of the early results were obtained using coincidence experiments with G-M counters, *not* cloud chambers as one might expect. This is because the counters could be arranged in many positions using various radiation absorbers like lead. This allowed the gathering of data from a variety of situations that would be extremely difficult with cloud chambers.

Possibly the first major result was obtained by Bruno Rossi in 1932. In an experiment, inspired by Skobeltzyn's description of multiple particle tracks observed in his magnet field cloud chamber, Rossi placed three Geiger counters in a *triangular* configuration, so that they could not be simultaneously discharged by a single particle traveling in a straight line.

With the trio of G-M counters enclosed in a lead box, he found about 35 coincidences per hour. With the top lead section *removed* the rate fell to nearly zero. Remarkably, it appeared as if more ionizing particles were being *generated* simultaneously with the top lead section *present*. How was that possible? There was nothing known at the time to explain this effect. The result was so astonishing that one journal refused to publish it, and another accepted it only after Werner Heisenberg vouched for Rossi's reliability.

Rossi, in 1933 discovered another important effect, now known as the *shower curve*. An illustration of this curve is shown in Figure 6.10 with Rossi's original data shown by small circles. The physical setup, depicted in the inset, consisted again of an array of three G-M counters arranged such that three *simultaneous* particles are required for a coincidence.

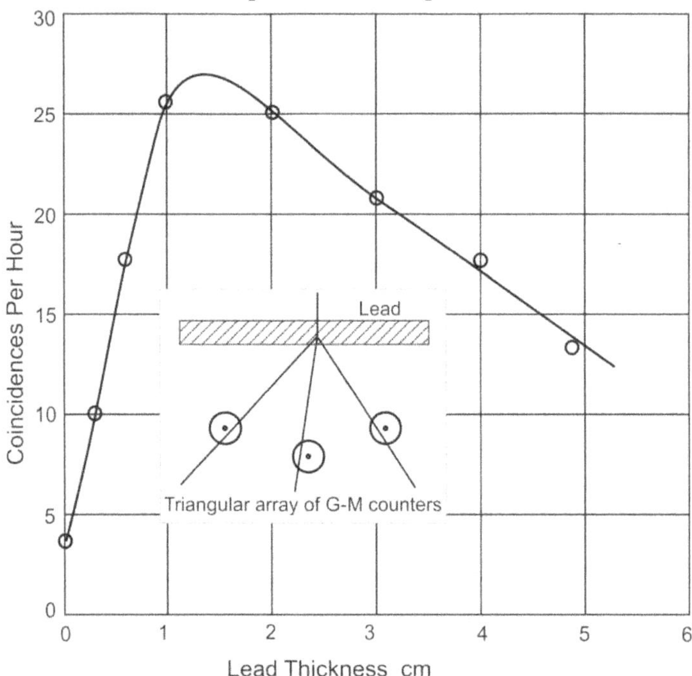

Figure 6.10 Rossi shower curve. It shows coincidence rate for lead of varying thickness. At least three charged particles emerging simultaneously from the lead are needed to produce a coincidence.

An examination of the curve shows that with *no* lead present, there are only about 4 coincidences per hour. But an amazing thing happens as lead is introduced *above* the G-M array. As the curve shows, when the lead thickness is increased, the rate of coincidences increases rapidly, due to showers coming out of the lead, reaching a maximum at 1.4 cm of lead. And then, unexpectedly, the rate decreases rapidly.

But this result was not consistent with what physicists thought they knew about the cosmic radiation. They assumed that the hard or penetrating particles found in the atmosphere initiated showers by interacting with atomic nuclei. But based on the shape of the Rossi curve this could not be the case.

Previous attenuation measurements had shown that the hard cosmic rays had a mean range in lead on the order of *meters*. But, in this case, the number of showers *decreased* when the lead thickness increased beyond 2 cm. Whatever radiation was causing showers was more easily absorbed by lead than by the penetrating particles found in the atmosphere.

In retrospect, the Rossi curve showed that the local cosmic rays consist of two components, a soft component capable of prolific generation of particle showers but rapidly attenuated in lead and the hard or penetrating component capable of traversing great thicknesses of lead and only occasionally giving rise to a shower.

Additional studies by Rossi and others with different materials were puzzling too. In these experiments, layers of lead, iron and aluminum, with the same *mass per unit* area, were placed above the triple counters. Based on current theory, it was expected that the materials would all attenuate by the same degree and produce the same showers. But amazingly this was not the case. The lead produced *two* times the showers of iron and *four* times as much as aluminum. Thus Rossi was forced to conclude that showers were produced much more abundantly in heavy elements than light elements. Other studies also showed that heavier elements were more effective absorbers of the radiation causing the showers.

It was now clear that Rossi had discovered a *major difference* between heavy and light materials for high energy particles. This peculiar feature was novel in physics. Up to this time it was thought that, for charged particles, ionization losses and Compton collisions were nearly the same in different materials with the same mass per unit area. Obviously this was not true for the new radiation.

Several conclusions could now be drawn from these observations of the shower producing radiation. An important theoretical picture was forming together with its place in the cosmic ray framework. As discussed in the next section, the theory developed from the skeleton of the existing theories.

6.4 Behavior of High Energy Electrons and Photons

Physicists thought they knew how *fast* electrons behaved in matter. For example when electrons are stopped instantly, photons are generated, usually as x-rays. But it was believed that only a very small amount of energy was lost by photon emission. The bulk of the energy was thought to be lost by ionization.

But then, in 1934, Hans Bethe and Walter Heitler published new results in an article titled "*On the Stopping of Fast Particles and on the Creation of Positive Electrons*", which pointed to a very different conclusion. They realized that a charged particle's trajectory becomes curved when it comes close to an atomic nucleus. This happens because of the strong electric field associated with the positive electric charge of the nucleus. Classical theories of electric and magnetic fields require that an accelerated charge emit electromagnetic waves. Since it is definitely true that motion along a curved trajectory is accelerated motion, the particle must radiate photons.

Bethe and Heitler used quantum electrodynamics (QED) to refine the classical physics results. QED, originally developed by Dirac, is essentially a combination of quantum theory and special relativity. It is used to describe how light and matter interact. The aim of their paper was to discuss in greater detail the rate of energy loss of fast particles by radiation and its dependence on the particle energy. Their findings were substantially different than those obtained with classical methods and caused quite a stir among physicists.

First, Bethe and Heitler found that radiation losses are much greater for light particles such as an electron than for heavy ones like a proton. This was explained as being due to acceleration differences. Since acceleration is inversely proportional to mass, a greater mass has less acceleration.

Secondly, radiation losses are greater in elements of high atomic number Z. This is explained by the fact that the deflecting force on a charged particle is proportional to the charge on the nucleus, i.e. the number of protons in the nucleus.

Third, the radiation loss will increase quickly with particle energy. For slower particles, ionization loss surely predominates. But it decreases quickly with increasing energy. Eventually, as the energy increases, the radiation losses increase and overtake ionization losses. For electrons this happens at about 10 MeV in lead and 100 MeV in air. This is shown in Figure 6.11. There is no counterpart to these findings in classical physics.

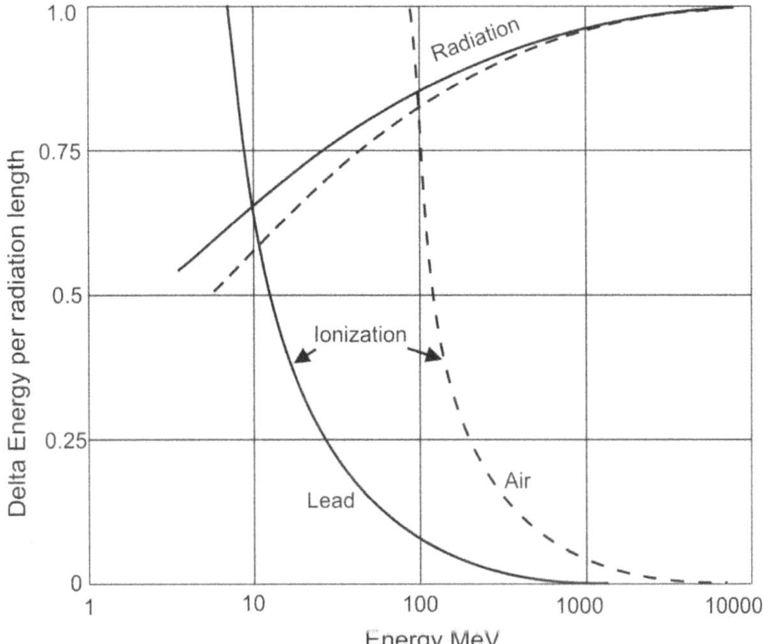

Figure 6.11 Fractional energy loss rate per radiation length for electrons through ionization and radiation processes. Data is normalized to particle radiation length.

Bethe and Heitler also presented some *startling* results on the creation of matter, known as pair production. Recall that Dirac predicted the creation of positive and negative electrons by photons. The most likely event is one where a high energy photon passes very close to an atomic nucleus.

Bethe and Heitler's theory predicted that, for a photon of given energy, pair production would increase quickly with atomic number. Like the radiation loss of electrons just discussed, it depends on the effect of the electric charge of the nuclei. Although photons have no charge, they are a form of electromagnetic radiation and will interact with an electric field and become scattered with a change in energy, the well known Compton Effect. Thus a photon, of given energy, has a higher probability of pair production near a high Z nucleus.

Their theory also predicted that above 1 MeV, where pair production begins, the probability of pair production quickly increases with photon energy, then levels off. It is known that Compton collisions decrease with increasing energy. Hence the low energy photons are absorbed mainly by the Compton Effect and higher energies by pair production. Pair production overtakes the Compton Effect at about 5 MeV in lead and 20 MeV in air. This can be seen in Figure 6.12.

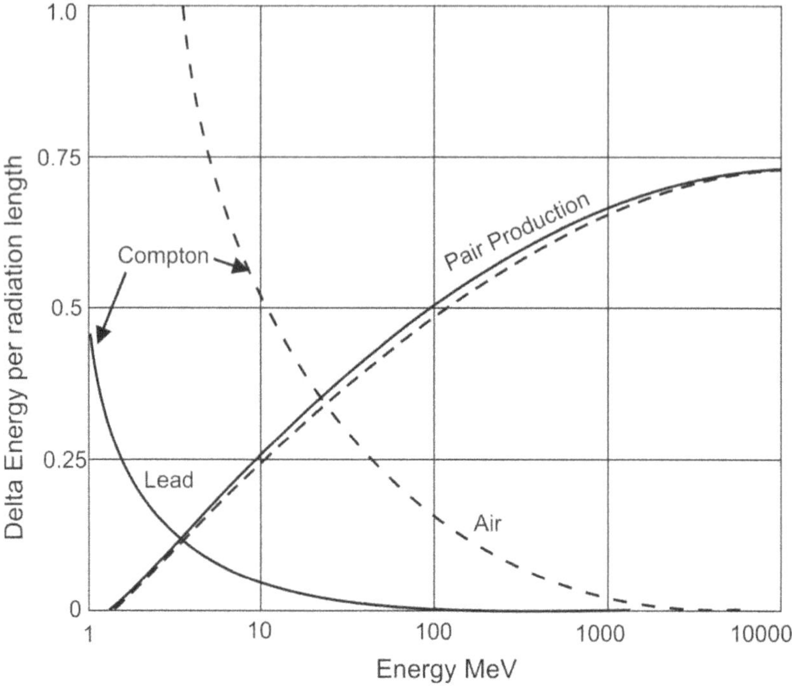

Figure 6.12 This shows the fractional energy loss rate, per radiation length, for photons through Compton collisions or pair production.

When Bethe and Heitler presented their theory it was believed that all particles in the local cosmic radiation were either electrons or positrons. But measurements by Rossi and others had shown that many of these particles could penetrate over a meter of lead. This was a stumbling block for the theory since it would have required electrons with ridiculously high energy capable of passing through such a thick absorber. Cloud chamber experiments of Anderson had shown that some charged particles reached 300 MeV and did not dissipate as much energy in matter as this theory predicted.

Faced with this evidence, Bethe and Heitler lost confidence in the theory. QED theory appeared to apply well for some cosmic rays and not for others. They thought that the absence of showers accompanying some of the fast particles indicated a breakdown of QED at high energies.

Sometime later, the particles that did not produce showers were identified as *mesons* and the evidence for a breakdown of the theory disappeared. The Bethe-Heitler calculation remained the gold standard for careful and accurate work in high-energy physics. In the next section we will examine how this theory unlocked the secret of cosmic ray showers.

6.5 Electromagnetic Cascade Showers

An explanation of cosmic ray showers was developed at about the same time in 1936 by Homi J. Bhabha and Heitler in England and J. F. Carlson and Oppenheimer in the United States. It depended upon the QED theory discussed in the previous section. Because the shower consists of only positive and negative electrons and photons, it is often referred to as an *electromagnetic* cascade. It can be described as a cascade or gradual building up of the shower through pair production and radiative collisions. It was very difficult to handle mathematically but is actually straightforward to picture. See Figure 6.13.

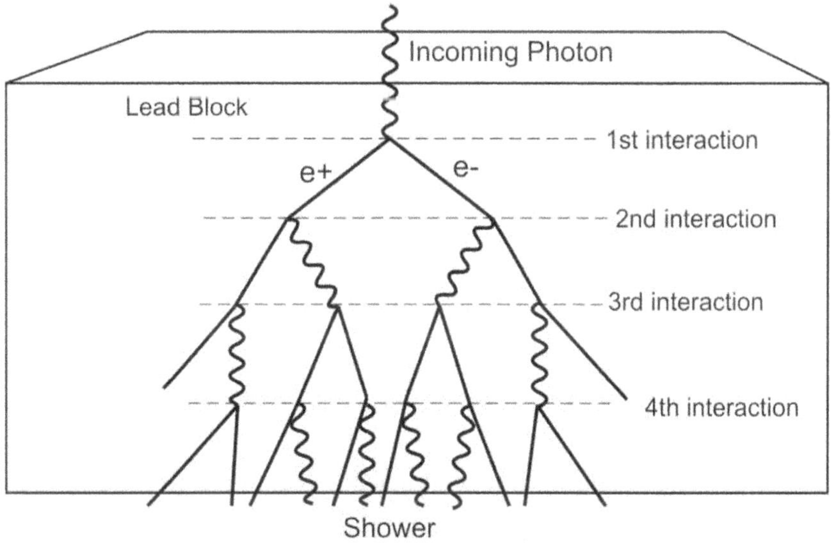

Figure 6.13 Development of a electromagnetic cascade shower in matter.

Say a high energy photon of 1000 MeV (1 GeV) falls on a block of lead. After travelling about 7 mm, based on the QED theory, it vanishes and is replaced by an electron and positron, which share the energy of the original photon. This is the first interaction in the figure. In another short distance, a second interaction, both charged particles radiate a photon losing a large part of its energy. The new photons soon materialize into an electron and positron and the process continues. At each new interaction two particles are produced from one, cutting the particle energy in half.

This process results in an increase of particles at first. But eventually the original energy is shared by more and more particles so that the new ones have little energy. Hence, the remaining photons are absorbed in Compton collisions, while the remaining electrons and positrons are brought to rest by ionization losses. Eventually the shower diminishes gradually and dies out. Figure 6.14 shows an actual photon induced shower as it develops through a series of lead plates in a large cloud chamber.

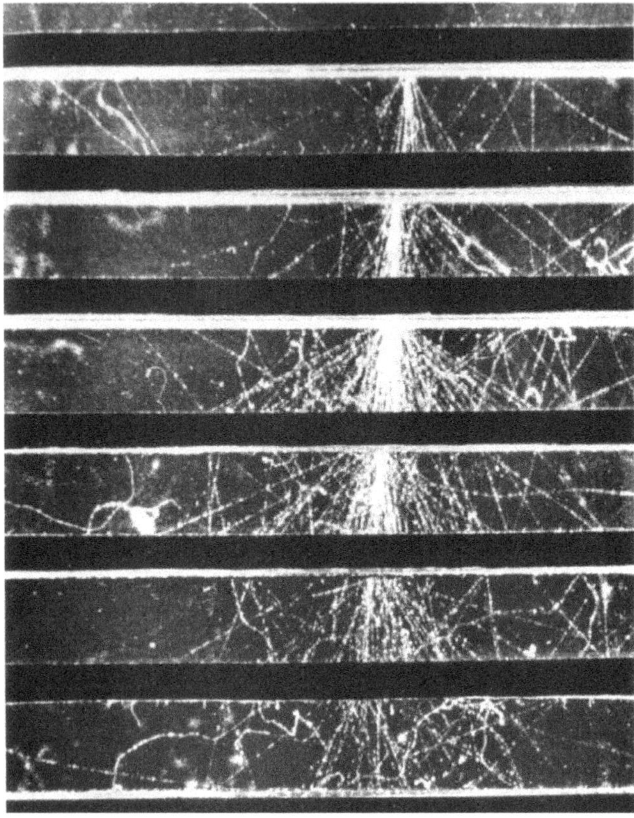

Figure 6.14 Photon initiated electromagnetic shower in a cloud chamber. The chamber contained 8 lead plates (only 7 shown here) 1.3 cm thick, placed across the chamber. The expansion of the chamber was triggered by coincident counters at sea level. The cascade was initiated by a primary photon of about 4 GeV. Some electrons are backscattered on the lead plates. (Courtesy cloudylabs.fr)

It is interesting to note that all showers are not alike. This is due to the statistical nature of the process. Precisely where a photon materializes in matter is a question of chance. Similarly, how a photon or electron shares the energy in the event is also a question of chance. Yet, the average behavior of showers can be predicted.

A cascade shower initiated by a high energy electron according to the shower theory is shown in Figure 6.15. Here it was assumed that a 1 GeV electron initiated the shower. It shows an amazing similarity to the experimental Rossi shower curve, shown in Section 6.2. The two curves do not represent exactly the same thing. Whereas the Rossi curve is based on recording coincidences in a trio of G-M counters, the theoretical curve shown here gives the number of electrons per thickness in lead. Never-the-less, this resemblance is not coincidental because the probability that a particle shower exits the lead increases with the number of particles in the shower.

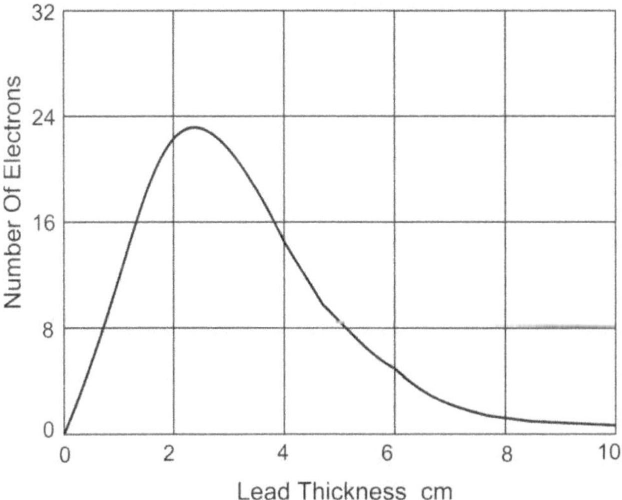

Figure 6.15 An electron produced electromagnetic shower in lead according to theory. The initiating particle is an electron of energy about 1 GeV (1000 MeV).

Many additional studies were performed which showed that the cascade theory explained the observed attributes seen in cloud chambers. Hence, theory and practice seemed to be in harmony. As an important consequence, the Bethe-Heitler theory was validated, except for its breakdown at high energies. Several other important facts were also established.

First, the sea level cosmic radiation contained, at the very least, electrons and photons with billions of electron volts. Secondly, the observed showers were the direct result of cascade processes started by these particles.

Third, the particle interactions in the cascade are mainly radiative collisions of electrons and pair production by photons. While these actions occur close to the atomic nuclei, they do not involve changes to the nuclear structure, as some had hypothesized.

Last, the ionizing particles that comprise the soft component of the sea level radiation are generated in the atmosphere and some in the material in the vicinity of the experiment.

The theory of electromagnetic cascades was a success. However, there was still a problem with the Bethe-Heitler theory which seemed to work well for showers and fall apart for high energies. The solution was not to come from any modification of the theory, but from the discovery of new particles.

6.6 Discovery of Muons in Cosmic Rays

The cascade theory of showers discussed in the previous section worked well for the *soft* component of cosmic rays. But it was becoming increasingly obvious that it did not work for the *hard* component.

For example, Neddermeyer and Anderson in 1934, using the same magnet cloud chamber in which the positron was discovered, found that most of the cosmic rays at sea level were highly penetrating. They could traverse large thicknesses of heavy material like lead and lose energy *only* by directly producing ionization. These particle tracks did not seem to ionize as heavily as might be expected from protons and less than expected for electrons. These observations did not fit the Bethe-Heitler theory.

Anderson remembers being perplexed by this paradox. "*They seemed to be neither electrons nor protons. We tended, however, to lean toward their interpretation as electrons and we resolved this paradox in our informal discussions by speaking of "green" and "red" electrons – the green electrons being the penetrating type and the red the absorbable type ...*"

This, then, was the situation in 1934 when the sea level penetrating particles had this paradoxical behavior. There was now a mystery to solve. What were these particles?

Two apparent scenarios were possibly. One is that the Bethe-Heitler theory is *valid* up to the highest energies but that the particles are *not* electrons. The other is that, if they are electrons, that the Bethe-Heitler theory *breaks down* for sufficiently high energies, several GeV, and their rate of loss is much smaller than predicted by theory. Both of these scenarios had their supporters.

The second scenario soon became indefensible on theoretical grounds. It was shown by E. J. Williams in 1934 that the Bethe-Heitler predictions must be valid well beyond energies of a few GeV. This presented a bit of a quandary as some of the observed showers were so large as to require particle energies of many GeV, while a large number of penetrating particles were found to have energies of one GeV or less. Hence, scientists began to consider the first scenario seriously.

If the new particles are not electrons, what could they be? At this point only two particles were known – the electron and proton. Hence, Williams and others suggested that the hard particle was a proton.

The proton hypothesis met with immediate difficulties. First, it would have implied the existence of negative protons since particles of both signs had been observed. Secondly, protons of 1 GeV are slow and ionize more than fast electrons. Observations in cloud chambers showed that the number of heavily ionized tracks is too small to be compatible with the proton hypotheses. More definitive answers were needed.

In 1935 Neddermeyer and Anderson began a series of crucial experiments, the first one on the summit of Pikes Peak. A number of cases of nuclear disintegrations were observed from which many protons were ejected but also showing, in a few cases, particle tracks lighter than protons and heavier than electrons. An example is shown in Figure 6.16. While this was not conclusive evidence of the existence of a new particle it did lend support to this assumption. In the next year, tests performed by the two experimenters brought forward very convincing evidence for this assumption.

As recalled by Anderson. "*In the summer of 1936, Neddermeyer and I were firmly convinced that all the data on cosmic rays, as known at that time, nearly forced on us the conclusion that the penetrating sea level particles could be neither electrons nor protons and must consist of a new particle type.*"

On November 12, 1936 Neddermeyer and Anderson presented evidence for the existence of new particles of *intermediate mass* at a colloquium at Caltech. A brief report also appeared in *Science*, November 20, 1936. The first formal publication related to the new particles appeared in the spring of 1937.

Before publishing their formal paper, they wished to obtain as convincing proof as possible that the theory was indeed valid at energies sufficiently high to require that the penetrating particles could not be electrons but must be particles of a new type. Normally, in their cloud chamber, they used a lead plate of about 3.5 mm thickness for studying energy loss in cosmic rays. Now, to get irrefutable evidence, they took an additional 6000 counter-tripped photographs in which they measured the energies of particles traversing a platinum plate of 1 cm in their cloud chamber. Since this plate was the equivalent of a whopping 1.96 cm of lead it reduced, to an insignificant level, the probability that an electron would pass through without producing a radiative collision.

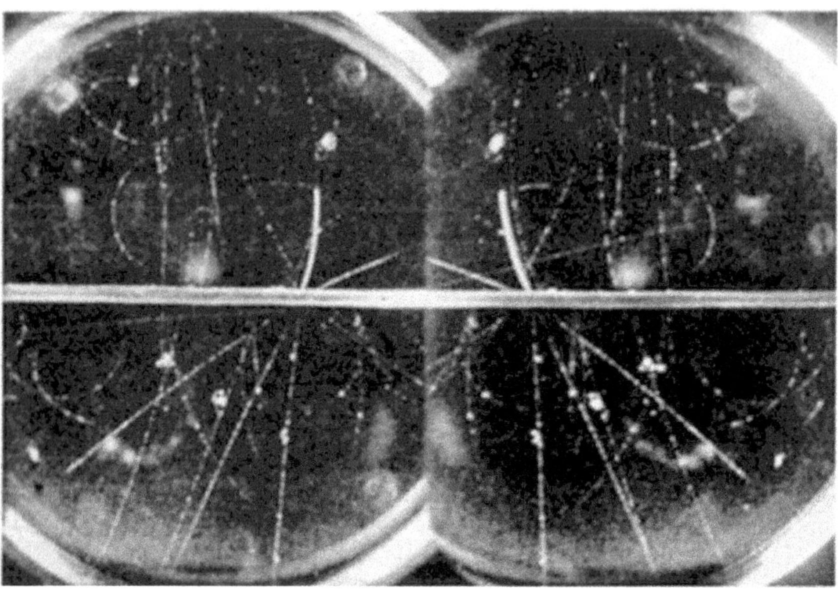

Figure 6.16 Pikes Peak cloud chamber 7900 gauss. A disintegration produced by a *non-ionizing* cosmic ray occurs at a point in the 0.35 cm lead plate. Six particles are ejected. One of the particles (strongly ionizing) is ejected vertically and has range of a 1.5 MeV proton. The other particle ejected upward to the right is probably an electron. The four particles ejected downward are probably positrons with energies of 105, 250, 500, and 60 MeV. The sum energy produced in this disintegration exceeds 1000 MeV. An unrelated electron shower is also seen coming in from above. The researchers concluded that the disintegration is fundamentally different than a shower. It shows that charge has been removed from the nucleus and made to appear in the form of light particles.

Shown in Figure 6.17 are the results of 55 measurements on the energy loss of particles in the range below 500 MeV. The ordinate shows the fractional energy loss of the particle in going through the platinum plate. Negative energy losses are due to errors in measurement. According to the researchers, "*It is clear that the particles separate themselves into two well defined groups, one consisting largely of shower particles, the other consisting of particles entering singly*". This was very convincing proof of their assumption. As a result of this paper the following facts were firmly established.

First, two very different groups of particles must exist. One group consists of penetrating particles which lose only a small amount of their energy traversing heavy material. Their energy loss is accounted for entirely by ionization. They do not radiate much. The other group consists of absorbable particles that lose their energies by radiation and ionization according to the theory. They are probably electrons.

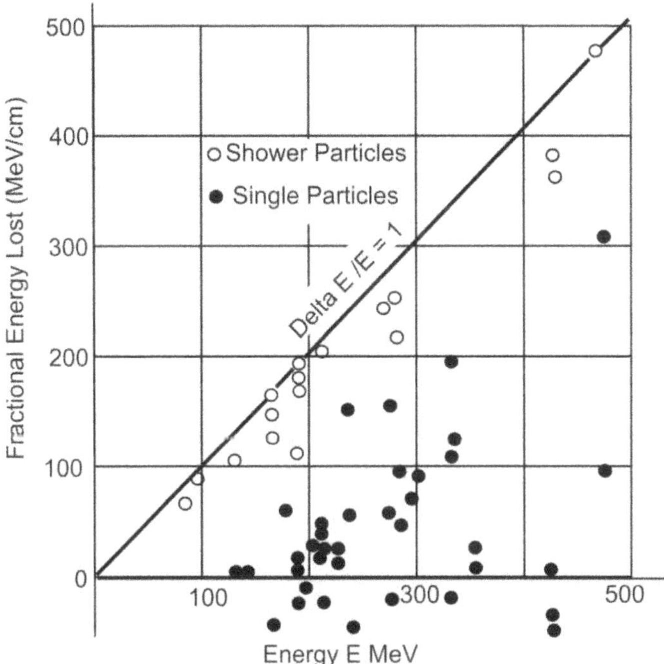

Figure 6.17 Energy losses of cosmic rays from data obtained by Neddermeyer and Anderson with 1 cm platinum plate in their cloud chamber.

Secondly, the penetrating particles usually appear as single tracks in cloud chamber pictures. Absorbable particles, however, are secondary particles of showers originating above the cloud chamber. Since these particles produce further showers, they are likely electrons.

Finally, since the penetrating particles ionize less than protons their mass must be *less* than the proton mass. And its mass must be *greater* than electrons since they do not radiate as much as electrons.

In their paper the two researchers said, "*that there exist particles of unit charge, but with a mass (which may not have a unique value) larger than that of a free electron and much smaller than that of a proton. This assumption would also account for the absence of numerous large radiative losses, as well as for the observed ionization.*"

At first, the new particle was known by various names, such as baryon, x-particle, Yukawa particle, meson, and heavy electron. One day Neddermeyer and Anderson sent off a note to *Nature* suggesting the name *mesoton.* When Millikan saw the note he reacted unfavorably suggesting the name should be *mesotron* since there was already the *electron* and *proton.* So the researchers sent off a cable to *Nature* asking them to reconsider, adding the letter "r". Fortunately, or not, the article appeared containing the name *mesotron.* Eventually *meson* won general acceptance and it became known as the mu meson or *muon.*

The discovery of the muon closed this phase of the investigation of cosmic rays. Particles in the local radiation were now known. The penetrating ones were muons and the absorbable ones were electrons. These two were responsible for most of the cosmic radiation on the ground. The next challenge was to find the mass and other properties of the muon. These studies are described in the next section. As is often the case, some unexpected properties were discovered along the way.

6.7 Properties of the Muon

At this point the *mass* of the new particle, the muon, was not determined. It was hypothesized in the previous section that the new particle's mass would be greater than that of an electron since they do not *radiate* as much as electrons. But this does not provide help in determining the mass, only of comparing relative masses. It was hoped that ionization methods might provide the answers.

In a magnetic cloud chamber the density of droplets gives the ionization density while the curvature of the track gives the magnetic rigidity. With these two quantities it is theoretically possible to calculate the mass. But measuring ionization densities accurately for high energy particles is very difficult.

Figure 6.18 shows relative (compared to electron) ionization densities for singly charged particles of varying masses intermediate between 100 times the electron and the proton. It is similar to Figure 3.8 but uses magnetic rigidity as abscissa rather than energy since that is what is used in cloud chambers. As seen, for magnetic rigidities above 10^6 gauss-cm it is very difficult to determine particle masses based on ionization density, as the ionization curves all come together at a low level. If the muons could be *slowed down* there would be better prospects for finding their mass.

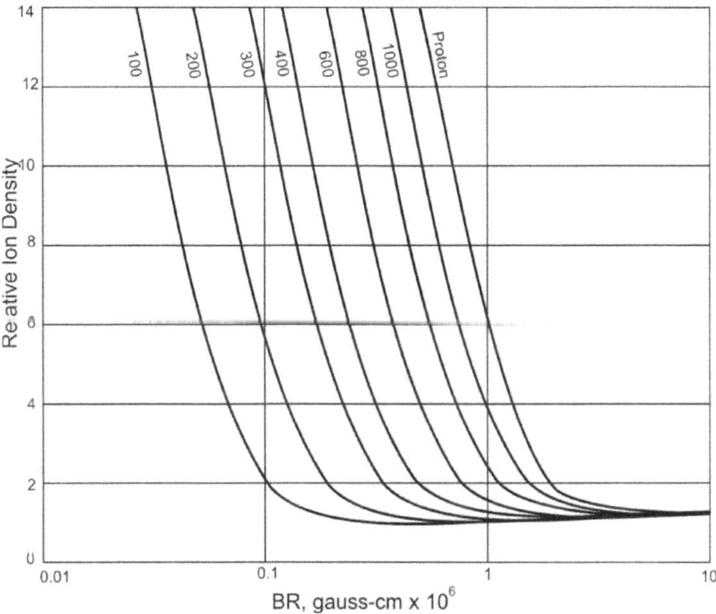

Figure 6.18 Relative ionization density, compared to electron, for particles of single charge with masses between 100 times the electron and the proton.

To illustrate, consider a particle track with a magnetic rigidity BR= 10^5 gauss-cm and relative ionization of six. This point falls close to the curve of 200 electron masses. Thus, if the particle is measured where the curves are widely separated good results can be obtained. Clearly, this method of measuring muon masses of low energy will work, but such low energy muons are exceedingly rare in practice.

In November, 1937, J. C. Street and E. C. Stevenson at Harvard decided to measure the muon mass by observing penetrating particles of *low* energy. Like other workers, they employed a cloud chamber in a magnetic field. However, to increase the yield of useful pictures they used an innovative system of lead blocks to *slow down* the particles, with G-M counters triggering photos only when particles stopped in the chamber. According to the workers, "*Tripping of the cloud chamber valve was delayed about one second to facilitate determination of the drop count along a track*". This allowed more precise measurement of particle ionization because of diffusion, but makes the track curvature less well defined.

Shown, in Figure 6.19, is an important photograph revealing a muon *stopping* in the chamber. It is seen coming out of the top of the region above the lead plate. This particle had a relative ionization of six compared to an electron and the curvature of the track indicated a magnetic rigidity of 9.6 x 10^4 gauss-cm. Street and Stevenson estimated it to have a rest mass of about 130 times that of an electron based on its velocity. This estimate is lower than one might obtain from Figure 6.18.

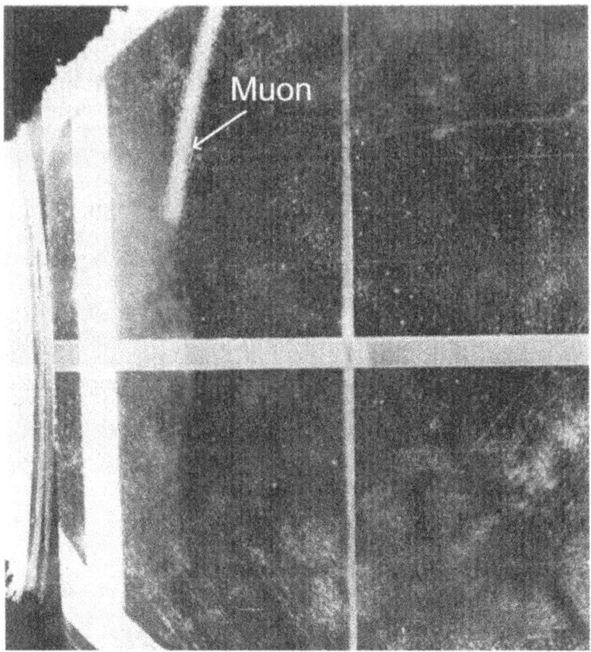

Figure 6.19 Mass of muon was found by Street and Stevenson from this cloud chamber photo of a particle track that had stopped in the chamber. It has a relative ionization density of six times that of an electron and BR= 9.5 x 10^4 gauss-cm.

A few months later, in June 1938, Neddermeyer and Anderson using their cloud chamber with 7900 gauss magnetic field observed a positively charged particle, estimated to be 240 electron masses and 10 MeV passing through a G-M tube counter inside the chamber and emerging with an energy of 0.21 MeV as shown in Figure 6.20. Interestingly, it comes to rest in the gas and disintegrates by emission of positive electron, not clear in photograph. They conclude, based on these considerations that the particle track cannot possibly be due to a particle of either electronic or protonic mass. When these results were published in *Physical Review*, they provided the most convincing proof of the muon existence. Additional measurements were made over the years to refine the value of the muon to 206.8 electron masses.

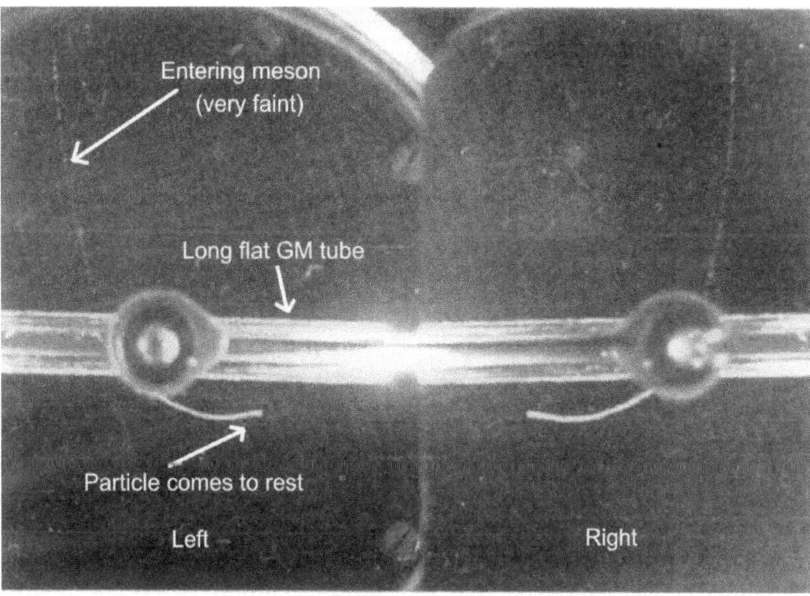

Figure 6.20 Stereographic cloud chamber photograph of particle decaying into meson by Neddermeyer and Anderson in 1938. This was the first convincing evidence for the existence of the meson (muon).

Quite soon after the announcement of the new particle Oppenheimer and Serber suggested that the new particle was the one postulated by Hideki Yukawa in 1935 to explain nuclear forces. The Japanese physicist had hypothesized the existence of a new particle several hundred times heavier than the electron. Yukawa used mathematics beyond the scope of this book but a simple explanation may help understand it.

It was known in 1935 that electric forces between the positive nucleus and the negative electron cloud are what hold an atom together, with the nucleus containing both protons and neutrons in a tiny volume. But the protons in the nucleus must *repel* each other since they have the same charge while the neutrons do not interact electrically. Since it is observed that most atoms are stable and the nucleus is holding together, there must be other forces at work, presumably on the quantum level, that is holding the nucleus together.

Yukawa assumed forces of attraction, of a *non-electric* nature, that held the nucleus together. These nuclear forces which acted over short distances, commensurate with the size of the nucleus, are much stronger than electric forces at small distances but weaker at large distances. He said that just as electric forces are associated with a photon, the nuclear forces should be associated with a particle. In studying the properties of this hypothetical particle he found that the nuclear forces required a particle of finite mass on the order of several hundred electron masses. More amazingly, he said the new particles, associated with the nuclear field, would be *unstable*. Here unstable meant the particles would decay into other particles in the same manner as radioactive materials. Moreover, the mean life of the particle would be on the order of one microsecond and the disintegration products would consist of one electron and one neutrino.

This hypothesis seemed to endorse the important theory of *beta decay* postulated by Wolfgang Pauli and Enrico Fermi wherein electrons are emitted in a variety of energies from radioactive elements. As a consequence, Yukawa's theory gained acceptance. But, as shown later in the book, Yukawa's meson had somewhat different properties than originally envisioned and was not the muon.

Nevertheless, Yukawa's meson, as it was called, was now *assumed* by many to be part of the cosmic ray flux and possibly the penetrating particles that were discovered at sea level. A number of researchers, guided by Yukawa's idea proposed that the particles might suffer decay while traversing a long path in air but not as much in thin, solid material. This might lead to a way to measure the decay. Hence the search was started to find the decay of the meson, if it existed.

Around 1937 research groups around the world began making *precise* measurements of cosmic ray intensity at various altitudes in the atmosphere. Initial results were rather confusing as they were contrary to Millikan's findings some years before. With the new measurements it appeared that air absorbed cosmic rays more effectively than solid or liquid matter with the same mass per unit area.

In 1938, a German physicist H. Kuhlenkampff proposed that if the particles were Yukawa's mesons, their decay would appear as an anomalous absorption of the mesons in air. The explanation for this is straightforward. Consider a layer of water 10 cm thick, equivalent to 8000 cm of air with the same mass per unit area. A meson travelling at nearly the speed of light will traverse the 10 cm of water in only a tiny fraction of a microsecond while the same meson travelling in air will need a large fraction of a microsecond. Hence, the meson with a mean life on the order of a microsecond will more likely undergo spontaneous decay while travelling through the air. Hence air will appear as a better absorber.

In the late thirties a number of scientists, among them Werner Heisenberg and H. Euler in Germany and Patrick Blackett in England studied the ramifications of Kuhlenkampff's idea. They also needed to consider the effects of temperature variations in air and Einstein's relativistic effect of time dilation. To complicate matters, at this time there was also support for the dualist theory of cosmic rays, wherein the electronic and penetrating components originate from different primary particles. This, indeed, was a very interesting time to be studying cosmic rays.

In September, 1938, Bruno Rossi (Fig 6.21) became interested in studying the decay of the muon. But the political climate in Italy was darkening with the enactment of Jewish laws. He was eventually deprived of his position at the University and soon thereafter left Italy. After brief stays at the Bohr Institute in Copenhagen and the University of Manchester he accepted an invitation from Arthur Compton at the University of Chicago.

Figure 6.21 Bruno Rossi (left) and Enrico Fermi at a meeting in the 1930's.

In the spring of 1939 he began "toying with the idea of testing the hypothesis of the instability of mesons by making a direct and precise comparison of meson absorption in air and in some dense material". Compton encouraged him and put at his disposal all the necessary equipment for the experiment. With two of Compton's assistants, Norman Hilberry and J. Barton Haig they set up a truck, a meson detector consisting of three G-M counters in coincidence, and enough lead surrounding the counters to remove the electron component.

They took measurements of cosmic ray counts at four locations of different altitude; Chicago (180 meters), Denver (1600 meters), Echo Lake (3240 meters) and Mount Evans (4300 meters). With the exception of Chicago, data was collected with and without a layer of graphite interposed above the counters.

Rossi found that the absorption in graphite was much less than in air. For example, at Echo Lake with an atmospheric depth of 699 g/cm^2, the addition of 87 g/cm^2 of graphite reduced the number of mesons by 10 per cent. But the addition of 87 g/cm^2 of air reduced the number by 20 per cent. The additional absorption through the air was due to the mesons decaying in flight. These measurements were enough to estimate roughly the mean life of mesons. Since the particles were travelling close to the speed of light, the calculations involved accounting for relativistic effects of time dilation and the particle's energy. This was possibly the first time that the time dilation effect was experimentally confirmed. After making suitable estimates for the particle's energy, they found a mean life for mesons at rest of about 2 microseconds.

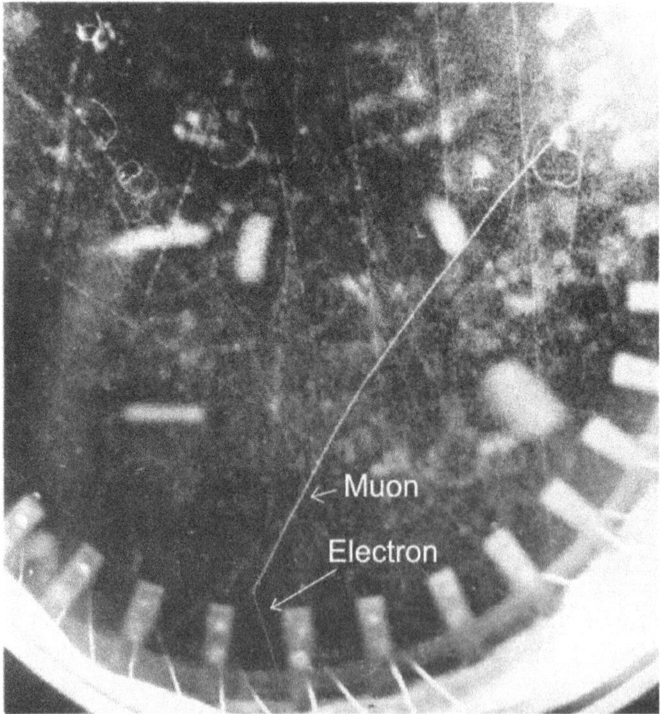

Figure 6.22 First cloud chamber photograph of a decaying positive muon reaching the end of its range in the chamber gas. A downward moving electron (very faint) is emitted at the end of the decay.

A short time later in 1940, E. J. Williams and G. E. Roberts in England obtained the first cloud chamber photograph of a decaying muon as shown in Figure 6.22. Their cloud chamber was 60 cm in diameter and 50 cm deep containing oxygen, with a magnetic field of 1180 gauss. No control system was used and only random tracks were photographed. Its large volume provided a long sensitive time and unbiased selection of tracks.

A slow muon is shown entering from the top right and stopping in the gas of the chamber about 5 cm before it reaches the chamber wall at the bottom. At this point a thinly ionizing decay electron appears. The researchers estimate the muon to be about 25 MeV with a mass of 250 electrons. The rectangular tabs in the photograph are supports for wire grids used by the electrostatic clearing field.

Measuring the mass of the muon and determining its decay properties did much to help understand the behavior of this particle. Its mass was found to be about 200 electron masses and it decayed in about 2 microseconds. Modern determinations of these findings are close to these values. It was also thought that the muon decayed into an electron and one or two neutrinos. The exact decay process was not clear at this time and required more study. Furthermore, some troubling reports seemed to indicate that the muon was not the meson that Yukawa had predicted. So, this part of the investigation was not over.

But the ominous tones of World War II were on the horizon. Thus, research into these puzzles would have to wait. Technology developed in the intervening war years would ultimately help uncover many new aspects of the cosmic radiation. They would include the discovery of an important new particle, the *pion*, responsible for nuclear forces that bind nucleons in atomic nuclei. These studies will be described in the next chapter where a more complete picture of cosmic rays will be developed.

Chapter 7

7 Cosmic Rays Before and After World War II

7.1 Research Environment for Cosmic Ray Studies after the War

During the Second World War most work on cosmic rays was justifiably postponed. When the conflict ended, scientists and others who were committed to the military effort now had to reflect on their future. Many of them were excited by the prospect of returning to academic positions they had left behind while others eagerly became students at those institutions. In the United States it was becoming increasing clear that a major change in science attitudes had occurred, with the focal point for research shifted to this country from Europe. This was true for several understandable reasons.

For one, the United States was spared damage because it was distant from the bombing that occurred in Europe and Asia. Hence its infrastructure was intact and considerable surplus scientific equipment was available. For another, the war effort had produced expertise in many fields such as radar, nuclear physics, and computer technology which served to establish a scientific preeminence. The government was eager to use this knowledge to make continuing advances in scientific research. And, finally, the US had become a magnet for scientific manpower from many war torn countries. Living conditions, especially in a ravaged Europe, were certainly not appealing after the war.

At that time there were big advances to absorb in many areas of science. For instance, in nuclear physics the experimental data was enormous and it was clear that further progress in this field would continue at a fast pace. There were also new developments in nuclear reactors and particle accelerating machines which promised to provide extended sources of man-made radiation for experimentation.

More surprisingly, science had captured the public imagination and it was willing to provide generous funding of science projects. Until the war, most scientific research in the United States had been supported by private foundations, local industries and undergraduate tuition fees. After the war, scientists experienced a continuity, and in some cases, an expansion of the wartime funding model.

Hence, the vast majority of all funding for basic research in the physical sciences in the United States came from defense oriented federal agencies, including the Department of Defense and the new Atomic Energy Commission, successor to the Manhattan Project. Also in the planning was the National Science Foundation (NSF) designed to promote the progress of science and basic research. It was later created as an independent federal agency in 1950.

Much of the work was conducted in institutions based on wartime models. Projects during the war had put together experts from many different fields of science and engineering to work towards shared goals, rather than grouping specialists by disciplines. Enormous time pressures and shared goals of war effort forced scientists and engineers to craft effective means of communicating with each other.

Leading scientists and administrators actively sought to continue the wartime spirit of collaboration between disciplines. The Atomic Energy Commission oversaw a new network of national laboratories to pursue both civilian and defense research. The labs featured interdisciplinary teams that mixed physicists, mathematicians and chemists with engineers of all disciplines. Similar set-ups appeared across dozens of US universities with facilities straddling several academic departments, such as the Laboratory for Nuclear Science and Engineering, founded at MIT by the end of 1945.

The facilities hummed with surplus equipment and know-how gained from the wartime projects. Physicist Bruno Rossi, for one, studied cosmic rays after the war by adapting the sensitive timing circuits he had built at Los Alamos to measure nuclear-fission rates.

Just months after the end of hostilities, physicists who had spent the war working on radar turned their attention to the impossibly small atom and the cosmically large universe. Some began to build radio telescopes and aimed them at the heavens. An international community coalesced, linking the Jodrell Bank telescopes near Manchester, UK, and the Parkes telescope in New South Wales, Australia, to similar instruments dotted across North America, from the California Institute of Technology in Pasadena to the National Radio Astronomy Observatory in Green Bank, West Virginia.

And in 1947, using repurposed microwave-frequency electronics left over from his wartime radar work, physicist Willis Lamb (Figure 7.1) of Columbia University in New York measured a tiny difference, of about one part in a million, between two energy levels of an electron in the orbit of a hydrogen atom. This effect was not explained by Dirac's formulation of quantum electrodynamics. Lamb's remarkable achievement challenged physicists' prevailing understanding of the vacuum, the mysterious state of lowest-possible energy.

In addition, the study of elementary particles was about to explode on the scene. Before the war only protons, neutrons, electrons, positrons, photons, neutrinos and a strange cosmic ray particle, the muon had been observed. The muon (called mesotron at this time) was unusual having a single charge and mass of about 200 electrons. The building blocks of the world looked relatively simple, though not as simple as in 1929 when only protons, electrons and photons were sufficient to explain the existence of matter. But some of the most informed physicists believed that other elementary particles existed and that it would be the next frontier in physics. Although hypothetical particles might present new theoretical problems, it was believed that modern instruments would play a role in their discovery.

Figure 7.1 Willis Eugene Lamb Jr. was an American physicist who won the Nobel Prize in Physics in 1955 "for his discoveries concerning the fine structure of the hydrogen spectrum."

Cosmic rays would lead the way in this quest until some years later when accelerators became widely available. Cosmic rays had energies higher than anything else in the laboratory at the time. They also had the distinct advantage of costing nothing, which was especially good for European nations devastated by the war. This allowed the revival of traditional cosmic ray studies in England and Italy immediately after the war.

This was the state of affairs in research institutions in the US and other parts of the world at this point in time. Many physicists considered their wartime work as an important but limited hiatus from their real work – scientific research. They were anxious to return to their original tasks. It wouldn't be too long before new discoveries concerning the atom and cosmic rays would appear. But first they had to resolve the difficulties concerning the current muon theory.

7.2 Doubts about the Current Muon Theory

In the 1930's scientists, searching for the solution of the cosmic ray mystery, found two previously unknown particles, the positron and the muon (called mesotron or meson at the time). These particles seemed to arise from the interactions of high energy cosmic rays with matter.

As is often the case in scientific research, progress leads to confusion, which leads to progress, which leads to confusion and so on. This has certainly been the case in the historical development of cosmic rays described so far. A similar situation arose with the discovery of the muon and the acceptance of Yukawa's mesotron as being this penetrating particle found at sea level. Here was the state of affairs.

Back in 1937, investigators like C. D. Anderson with his collaborator, S. H. Neddermeyer, as well as M. L. Stevenson, R. B. Brode and others found cosmic rays of an intermediate mass between that of the electron and the proton. Their measurements showed that these particles had a mass of about 200 electrons, are unstable and decay with a mean life of about 2 microseconds.

As mentioned in the previous chapter, Oppenheimer and Serber suggested that the new particle was the one postulated by Hideki Yukawa (Figure 7.2) in 1935 to explain nuclear forces. The Japanese physicist had hypothesized the existence of a new particle several hundred times heavier than the electron. This hypothesis seemed to endorse the important theory of *beta decay* postulated by Wolfgang Pauli and Enrico Fermi wherein electrons are emitted in a variety of energies from radioactive elements. As a consequence, Yukawa's theory gained acceptance.

The success of Yukawa's theory stirred a lot of excitement, but an intensive look at the facts should have raised some doubts. These new particles were supposed to be the glue holding all the protons and neutrons together in the nucleus. If so, they should be absorbed, shortly after their formation, by nuclei of atoms in the atmosphere. And yet, their flux at sea level and even at lake bottoms was not appreciably different from their flux on mountain tops.

In 1938 Yukawa and Sakata published a more detailed version of the theory. This time they predicted that the Yukawa particle would have a lifetime of one-hundredth of a microsecond, about 200 times less than current measurements by Rossi and others. Various attempts were made by Yukawa to get rid of the discrepancy, but it could not be done. Hence, doubts were cast on this particle being the one observed by Anderson and others.

In 1940 Sin-itiro Tomonaga and Gentaro Araki in Japan pointed out that positive and negative mesotrons should behave in a radically different manner after coming to rest in matter. They argued that a *positive mesotron* can never approach very near an atomic nucleus because of the electric forces of repulsion exerted by the positive nuclear charge. Hence it will simply wait around between atoms until it disappears by spontaneous decay.

Figure 7.2 Hideki Yukawa giving a talk at Columbia University in 1949.

On the other hand, the *negative mesotron* is attracted to the nucleus and might be captured before it has a chance to escape. If the interaction of the negative mesotron and the absorbing nucleus is sufficiently strong almost all negative mesotrons would be captured into atomic-like orbits, with very small radii. Consequently, they would overlap the nucleus substantially. Given that the Yukawa particle was designed to explain nuclear forces, it would certainly interact extremely rapidly with the nucleus, being absorbed long before it could decay directly.

Therefore, the life expectancy of negative mesotrons must be less than positive ones, since positive ones die a natural death while the negative ones are exposed to additional risk of accidental death by nuclear capture. When captured, the negative mesotrons are absorbed with their mass turning into energy. This sudden release of energy should cause the nucleus to explode.

In 1941 Franco Rasetti made an important step. By comparing the number of mesotrons stopping in an iron block and observing the number of decay electrons coming out of it he was able to show that only half the mesotrons decayed into electrons. The natural conclusion was that the decay electrons came from positive mesotrons while the negative mesotrons disappeared by nuclear capture as hypothesized by Tomonaga and Araki.

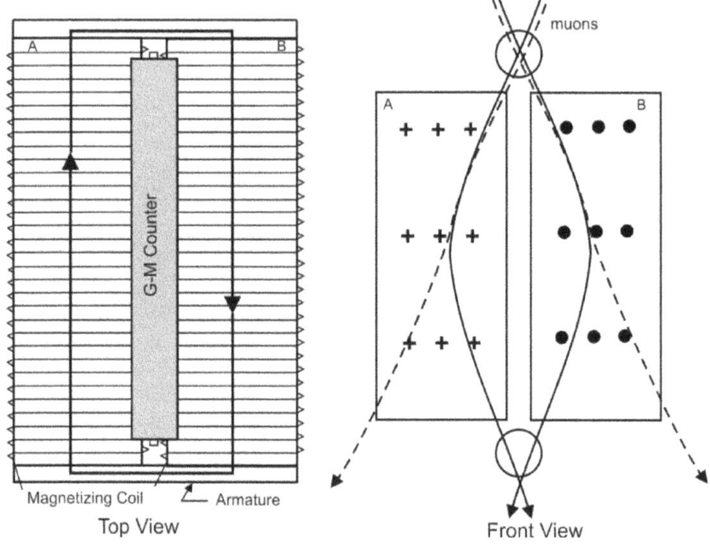

Figure 7.3 Magnetic lens devised by Rossi in 1931 for his measurement of magnetic deviation of muons. Magnetic circuit is shown by arrows in top view.

Focusing on the Tomonaga effect, more conclusive evidence was obtained by M. Conversi, E. Pancini, and O. Piccioni working under very difficult circumstances in Italy during the war. They employed a novel method of separating positive and negative mesotrons for measurement - the *magnetic lens*. This device was invented by Rossi in 1931 in an attempt to find out whether cosmic ray particles capable of traversing large thicknesses of matter were positively charged or negatively charged. His experimental results were ambiguous. He could only conclude that the particles had too much energy to undergo a measureable magnetic deflection in the iron of the lens or that positive and negative particles were equally abundant. Rossi's magnetic lens is shown in Figure 7.3.

As seen in the top view, the core of Rossi's lens consisted of two easily magnetized iron blocks A and B. Armatures at each end closed the magnetic circuit. The wire carrying the magnetizing current was wound round the blocks in a single layer; closed induction lines pass through the core, as indicated by the arrows, clockwise or counter-clockwise, according to the direction of the magnetizing current. Above and below the magnet are two G-M counters; their axes are horizontal and parallel to the direction of magnetization.

As seen in the front view, particles passing through the upper counter are concentrated toward the lower counter or deflected away from it depending on the sign of their charge. Shown are trajectories of positive muons (solid) and negative muons (dashed). For opposite magnetizing current the situation would be reversed.

Conversi and his colleagues employed the magnetic lens, essentially the same one designed by Rossi, to select either positive or negative penetrating particles from cosmic rays and then determine whether they decayed or not when stopped in matter. A simplified sketch of their setup is shown in Figure 7.4.

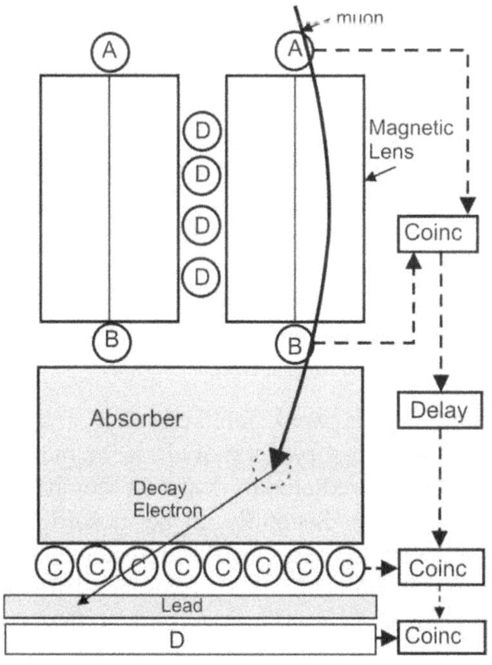

Figure 7.4 Layout of experiment by Marcello Conversi and colleagues

It worked this way. A mesotron (muon) of either positive or negative charge enters from above. Now assume the lens magnetization is set to focus positive particles. A positive muon may hit G-M counter A and curve downward as shown in Figure 7.4. Coincidence between counters A and B indicate that it is a mesotron. After spending some time in the absorber it will decay, producing an electron, which can be detected by one of the G-M counters C. If C is coincidence with A and B then it must have been a decaying particle.

On the other hand if the entering particle is not positive it will be rejected by the magnetic lens. By reversing the magnetization in the lens a similar procedure can be done to examine negative particles. The D counters provide a veto of the count for the case of random radiation.

Conversi and his colleagues found that positive and negative mesotrons stopping in matter behave differently. The positive ones decay more or less as though in a vacuum. However, the negative ones, if stopped by *heavy* nuclei such as iron, will be captured and produce disintegrations. But when the negative ones are captured by *light* nuclei such as carbon (Z=6) a sizeable fraction of them still decay as though in a vacuum.

It is not surprising that nuclear capture increases with increasing atomic number. A heavy nucleus has a greater positive charge so it would have a stronger pull for negative mesotrons. But some scientists found it hard to understand how negative mesotrons could evade capture in a light element, such as carbon. Their line of reasoning was based on the measured mean life of about 2 microseconds for the mesotron. This mean life was millions of times longer than the most generous estimates for nuclear capture. Thus, on theoretical grounds it was clear that all negative mesotrons stopped in carbon would be captured almost immediately before they had a chance to decay. If, as the data showed, most negative mesotrons did undergo spontaneous decay, they must have retained their identity *within* the nuclei for a time period great enough that they had a chance to come out again. In physics this is called a *weak* interaction.

Scientists found such a weak interaction, if it existed, as not being consistent with Yukawa's theory. First of all, if the mesotrons were the glue of the nuclear force field predicted by Yukawa they had to interact strongly with protons and neutrons. Secondly, if the mesotrons were produced by the passage of high energy particles through nuclei in the atmosphere it would take less than a millionth of a microsecond for this to happen. But the experiments showed that mesotrons took a few microseconds to be absorbed.

Therefore, the Italians concluded that, contrary to Yukawa’s theory, the interactivity of mesotrons with nuclear particles might vary with nuclear charge for some unclear reason. Citing this work, E. Fermi, E. Teller and V. Weisskopf showed that the decay of negative mesotrons in matter was twelve powers of ten longer than the time needed for a Yukawa particle to be captured by a nucleus via the mechanism of the nuclear forces. Hence there were now most serious doubts to this point that the mesotron was Yukawa's particle.

In 1947, a number of physicists including Tanikawa, Sakata, and Inoue in Japan as well as H. Bethe and R. Marshak in the United States independently put forward a hypothesis that would have removed the difficulties. They proposed that the observed mesotrons were the decay product of Yukawa's mesotron, which no one has yet observed. This attractive and plausible hypothesis had yet to be proved.

The answer to this puzzle came a few years later with the perfection of a new detector – the nuclear emulsion. This occurred in Vienna and the University of Bristol in England.

7.2 Photographic Emulsions Come Of Age

Most of the tools used by physicists at this time were truly superb instruments. Perhaps, the most complicated and useful of these was the cloud chamber. It was described by Ernest Rutherford as, "*the most original and wonderful instrument in scientific history*". In the 1930s in particular, the cloud chamber proved to be an invaluable tool for investigating cosmic rays. Being a visual device, the cloud chamber enabled scientists to see charged particle trails and take their picture. The first muon was seen in this way by the young scientist, Dimitrij Skobeltzyn, in Leningrad in 1928.

However, these devices have their limitations when studying particle interactions. For instance, cloud chambers ordinarily use a low density gas and consequently only a few particles that enter the chamber actually interact with the gas. A significant improvement occurred when a thin lead plate was inserted into the chamber. This provided a heavier mass for particle interaction. Carl D. Anderson used such an assembly to discover the positron in 1932.

As scientists built larger chambers they used many layers of materials which proved very helpful in seeing multiple particle interactions. These interactions with the plates were valuable, but because of the thickness of the plates it was impossible to see exactly what happens at the collision point. Also, if a particle stops in the plate and decays, the resulting products can only be observed if they have enough energy to get out of the plate.

Thus it was clear that seeing interactions and decay processes requires using dense substances in which the particles have a good probability of colliding and coming to rest while also leaving visible tracks in the chamber. Such conflicting requirements were a major impediment to seeing tiny particle interactions at this time. Photo emulsions held promise of an alternate pathway to this goal.

A photographic emulsion was first used in nuclear physics research in 1896 by A. H. Becquerel, who detected the radioactivity of uranium salts, from the fogging the salts caused on a emulsion. In 1910 the Japanese physicist S. Kinoshita demonstrated that silver halide grains in an ordinary photographic emulsion become capable of development if even a single alpha particle passes through them. Researchers soon became aware that photographic emulsions, with their long exposure times, had a great advantage for recording rare events.

What may have been the first use of photographic emulsions on a balloon flight occurred in November 11, 1933 when the Explorer II lifted off from the Stratobowl in the Black Hills of South Dakota. It carried Captain Albert Stevens and Captain Orvil Anderson of the U.S. Army and many scientific experiments including an array of Geiger Müller tubes and Kodak film plates. Despite some problems with the equipment they observed the number of cosmic rays entering the vertical telescope increased steadily up to 57,000 feet where it peaked at 55 times greater than sea level. Dr. T. R. Wilkins of the University of Rochester provided the two packages of film plates, wrapped in lightproof black paper, that were attached to the outside of the gondola. After the Explorer II flight, Wilkins discovered a track on the plates, probably made by an alpha particle, with an energy level of 100 million electron volts.

About ten years earlier in 1923 Marietta Blau (Figure 7.5), at the Institute for Radium Research in Vienna, began her seminal contributions to every conceivable aspect of emulsion physics. Particle tracks on film held the promise of reproducibility and objectivity, but early emulsions were notoriously fragile and unreliable. She studied emulsions of varying sensitivity to visible light as well as to alpha particles, fast protons and gamma radiation. Using commercial emulsions, she successfully distinguished alpha particle tracks from fast protons and background radiation. She also determined proton energies by measuring distances between exposed grains in their tracks.

Beginning in 1932, Blau and her student Hertha Wambacher measured the energies of recoil protons in hydrogen rich emulsions. At this time she also persuaded British film manufacturer Ilford to thicken the emulsion on its film and experimented with every other parameter; grain size, image retention, and other factors to improve visibility of tracks.

In the next years the emulsions improved and they refined their development. Then in 1936 Ilford made available exceptionally thick emulsions. Blau and Wambacher saw an opportunity to record long tracks of extraterrestrial protons using this new emulsion. They chose to use the research station operated by Victor Hess on the Hafelekar Mountain near Innsbruck at 2300 meters above sea level. They installed a vertical stack of plates and exposed them for four months. Blau and Wambacher recorded the expected particles, protons and neutrons but to their amazement they also found four cases of "*stars*", each with several heavy tracks– from three to twelve – radiating from a central point. Nothing like it had been seen before.

Figure 7.5 Marietta Blau in her laboratory in Vienna at the Institute for Radium Research about the year 1925. A film stack and microscope are on the table.

Shown in Figure 7.6 is one of four 'stars' observed by Blau and Wambacher using a 70 micrometer thick emulsion layer. It was not possible to focus all the tracks simultaneously because of the steep angle that the particles made crossing the emulsion. The center of the star is 25 micrometers under the surface of the emulsion.

They quickly sent off a note to Nature stating that the stars were due to the disintegration of heavy nuclei (such as silver or bromine) in the emulsion and that they could only have been caused by high energy cosmic particles. These stars with their abundance of heavy tracks provided the first indisputable evidence that the disintegration of heavy nuclei was real. This discovery created extraordinary excitement in the cosmic ray community.

For one brief moment Blau was at the peak of her career. But the golden moment ended sharply with the deteriorating political situation in Austria in 1938. Shortly after her 'star' discovery she found it necessary to flee Vienna, first to Oslo and finally to Mexico City where she became a professor of physics at the Polytechnic School until 1944. Then Blau moved north to New York City where she worked at various places for the duration of the war.

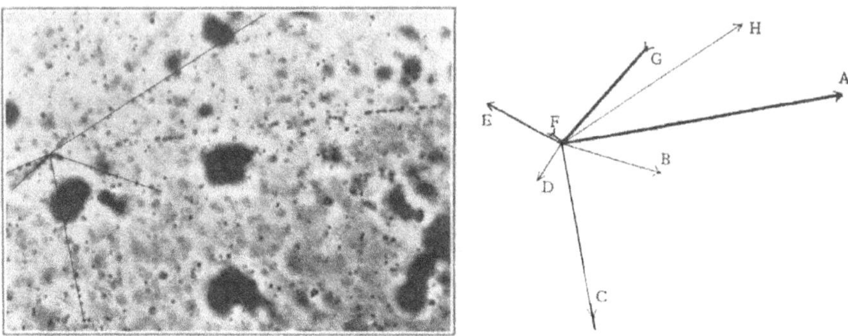

Figure 7.6 Photographic plate exposed to cosmic radiation on Hafelekar Mountain by Blau and Wambacher showing a 'star'. Left side shows original image with fine lines added to help see the particle tracks. On the right is a sketch of the same star made by the two Austrians.

In 1938, C. F. Powell turned from cloud chambers to photographic methods after he and others in the field understood the importance of Blau and Wambacher's use of emulsions for studying cosmic rays. During the war Powell organized the photographic technique into a large enterprise with others and made significant discoveries as we will see later.

In the 1930's manufacturing of photographic emulsions with standardized properties began, but they were used mainly to register the tracks of slow particles, mainly alpha particles and protons. Things continued to improve and by the middle 1940's scientists were able to create an improved detector for cosmic rays – the nuclear emulsion. Its operation is similar to a photographic emulsion.

When high energy charged particles hit a photographic emulsion they produce the same result as being exposed to light. Effectively, these particles cause a change in the grains of silver bromide that they see along their path in the emulsion. When developed with an appropriate liquid solution, the sensitized grains will be reduced to silver. Observing the grains with a microscope, the paths of the individual particles can be seen as rows of dark silver specks.

However, the first photographic emulsions were sensitive only to slow particles that leave a dense trail of ions along their path. Until the mid 1940's standardized emulsions were very thin and only particles moving nearly parallel to the emulsion plate had a chance of being registered. Because of this, photographic emulsions were seldom considered as a method for detecting charged particles.

At the University of Bristol in England under the leadership of C. F. Powell and G. P. Occhialini the emulsion technique had a renaissance. In collaboration with the Ilford Company they prepared novel emulsions that contained a higher concentration of silver bromide than standard photographic film, which turned out to be more sensitive to ionizing particles. Soon thereafter the Kodak Company in the US began working in this field.

There are two features that distinguish nuclear emulsions from ordinary photographic emulsions. First, the mass ratio silver halide to gelatin is eight times higher. Second, the emulsion thickness is ten to one hundred times greater, reaching 1 mm or more.

As mentioned earlier, charged particles or electromagnetic radiation associated with nuclear reactions affect nuclear emulsions in a manner similar to light. It was also found that the process of development greatly *amplifies* the initial weak latent image, similar to the way the avalanche in a Geiger counter chamber greatly magnifies the weak effects associated with the initial ionization produced by a charged particle.

Emulsions used in nuclear physics are usually layers applied to glass supports. In the study of high-energy particles in cosmic radiation, the emulsions are often stacked in several hundred layers. The volume of the stacks reaches tens of liters to form a nearly solid photosensitive mass. After exposure, individual layers may be cemented to glass supports and processed in the conventional manner. The position of the layers is precisely marked so that the trajectory of particles passing from layer to layer can be easily traced through the entire stack. A later development was the stripped emulsion that did not use glass backing and stacked into layers to form an emulsion chamber.

Researchers soon became skilled in recognizing properties of the particles from their tracks in the emulsions. For instance, the density of black silver grains on a track corresponds to the density of ion pairs produced in a gas. Thus, as in cloud chambers, the density decreases with increasing velocity in a well known manner. These and other measurement tricks were useful in determining the mass and range of the particle.

Things were working out well and the technology of nuclear emulsion plates was ready to be tried out at several mountain altitudes. In the next section we will relate some of the discoveries from these experiments and their implications to cosmic rays.

7.3 Discovery of the Pion

Nuclear emulsion plates had come of age in the 1940's. But it took a lot of effort to make it happen. As noted in the previous section, early photographic emulsions had a number of drawbacks. For one thing, they were required to be very thin, a fraction of a millimeter thick, in order to be developed properly. Hence only particles travelling within the narrow plane of the emulsion would be registered. Those passing perpendicular to the emulsion would leave only a spot. Also, the early emulsions were only sensitive to slow particles. Fast ones would pass through without leaving a trace. With the help of Ilford and later on Kodak, the nuclear emulsion became thicker and more sensitive, thus becoming a useful tool for investigating cosmic rays.

Late in 1946, C. M. Lattes, G. P. Occhialini and C. F. Powell, began using the newly formulated Ilford C2 nuclear plates at mountain heights. Occhialini took a stack of two dozen 2 cm x 1 cm plates coated with the latest silver bromide emulsions to Pic du Midi, at 2800 m in the French Pyrenees for exposure at a cosmic ray observatory. After a few weeks exposure, the plates were developed and 65 mesotron (muon) tracks were found. They also found several peculiar tracks of a particle entering the emulsion and coming to rest and giving rise to a second particle.

A photomicrograph of this two-meson decay is shown in Figure 7.7. The tracks gave such convincing evidence for two-meson decay that Lattes, Muirhead, Occhialini and Powell published their findings in *Nature* in May 1947. Aware of the problems of the identification of the cosmic ray mesotron with the Yukawa nuclear force meson, they wrote "*Since our observations indicate a new mode of decay of mesons, it is possible that they may contribute to a solution of these difficulties*". Clearly, more evidence was needed to justify such a far-reaching conclusion.

But, for some time thereafter no more two-meson decay tracks could be found at Pic du Midi. At this time Lattes suggested going to Mount Chakaltaya at the height of 5500 m in Bolivia Andes, near the capital La Paz where a weather station was located. After exposing the plates on the mountain for one month they were returned to Bristol. Here they found 644 tracks of mesotrons and also ten more tracks of a particle stopping and decaying in the emulsion. With the Bolivia plates, the Bristol group had the confirmation they needed.

By studying the photomicrographs the Bristol researchers discovered many features of the particles. A gradual increase of the density of the grains along the track clearly shows the direction the particle is travelling. This is deduced from the fact that particles slowing down will ionize more heavily. The rate, at which the density increases, as well as comparisons to proton tracks in similar emulsions, indicates the mass. In this case they found a mass of several hundred electron masses.

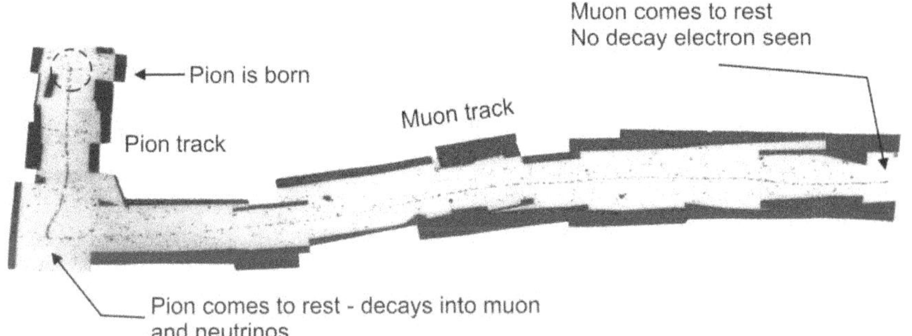

Figure 7.7 One of the first photomicrographs of a pion to muon decay from the Bristol group in England using a new Ilford emulsion. This emulsion is not sensitive enough to record the decay electron at the end of the muon track.

When the first particle comes to rest, another particle emerges as seen in Figure 7.7. This second particle also comes to rest in the emulsion after travelling about 0.5 mm. It appears to also have a mass of several hundred electron masses. But, very careful measurements of the grain densities, at various places along the track, show conclusively that the second particle had slightly less mass than the first one.

Now the researchers had to reflect on other possibilities that might have caused the unusual tracks. After considering several alternative explanations, Lattes, Occhialini and Powell came to the only logical conclusion – one mesotron had decayed into another one. These results from the Bolivian Andes were presented in the journal Nature in October 1947. Here they stated that they had found forty examples of the process leading to the production of secondary mesons. This result was conclusive, but its interpretation presented some problems.

Clearly, the heavier (parent) particle could not be the mesotron associated with the penetrating part of the cosmic radiation as it had already been shown that when that mesotron decays, an electron is produced not another mesotron. But the tracks would make sense if there were at least two kinds of mesotrons, with the lighter (daughter) particle being associated with the penetrating part of the cosmic radiation. But, for this explanation to be correct, the secondary mesotron, after coming to rest, should decay and produce an electron. But this was not seen on the emulsion.

Might it be possible that the decay electron could not be seen on the emulsion because it was not sensitive enough to register an electron? It turns out that this was precisely the case; for when Kodak developed more sensitive emulsions in 1948, a decay electron could be seen as shown Figure 7.8.

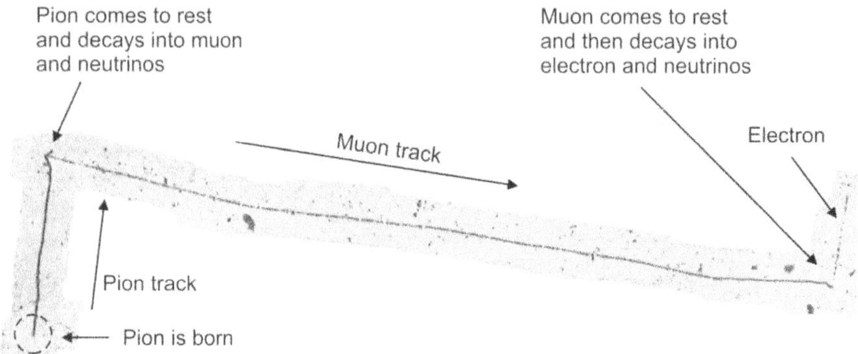

Figure 7.8 This is a photomicrograph from the Bristol group of a pion coming to rest and decaying into a muon. In turn, the muon comes to rest and decays into an electron.

There was no longer any doubt that the Bristol group had discovered a new mesotron. The heavier parent particle became known as the π meson (now known as the pion). The lighter secondary particle, which was originally known as Anderson's mesotron, was then named the μ meson (now known as the muon).

The puzzle created by the discovery of the original muon was substantially cleared up by the arrival of this new particle. The nuclear force "*quanta*", predicted by Yukawa, were not muons but pions. It was the pions that were produced in high energy impacts associated with the nuclei of atoms. It was no longer surprising that muons did not interact strongly with nuclei since they were not directly produced in such nuclear reactions. They were born of the decay of pions. This was a different picture from the one presented by Yukawa where the new born particle would decay into an electron.

This new picture required some explanation concerning the disposition of the pions in matter. If pions were the particles of Yukawa's theory and were abundantly produced in high energy interactions, then the *negative* pions should be promptly captured in matter. This capture should produce a sudden release of energy in effect causing a nuclear explosion. As it turns out that is precisely what happens.

An examination of many tracks in nuclear emulsions showed that negative pions were never seen to decay into muons. Instead, a *star* is produced at the end of the track as seen in Figure 7.9. This was first shown by D. H. Perkins, Occhialini and Powell in 1947. The star shows many of the particle fragments released as a result of the explosion. Thus, the negative pion that stops in matter has no chance to decay before being trapped by the nucleus and exploding.

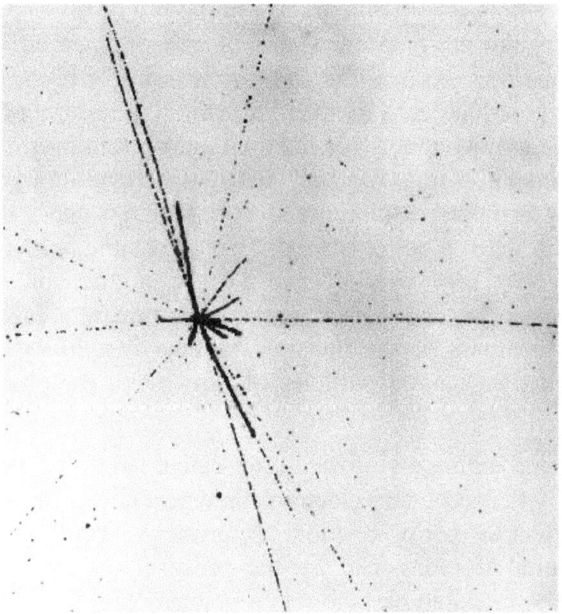

Figure 7.9 Disintegration of a silver nucleus by a cosmic ray particle. Energy of the particle is estimated to be 1000 MeV. Tracks of seven protons, five alpha particles and some heavier nuclear fragments can be distinguished.

There are now two unmistakable signatures given off when a pion stops in an emulsion. A positive pion decays into a muon and then an electron. While a negative pion is captured by the nuclei forming an exploding star.

What about pions at sea level? Do any of the pions survive the journey through the atmosphere to the Earth's surface? The answer depends on the mean lifetime of the pion. Early calculations indicated a mean life very much shorter than a microsecond. A more precise value of mean life, using high-energy accelerators in 1948, was found to be 2.55 x 10^{-8} seconds. This is about 100 times shorter than a muon. Thus pions would not travel as far as muons because they decay faster and readily interact with atomic nuclei. Consequently, the sea level penetrating radiation consists almost entirely of muons, not pions.

There was one more question to be answered. What are the

disintegration components of the decay of these particles? Yukawa's theory predicted that his nuclear force particle would decay into an electron and a neutrino. But this was not the case since it was seen that pions decayed into muons. Hence there was not a reliable theory available to predict what happens in this situation. So it was necessary for experimenters to lead the way.

First consider the decay of pions to muons. Recall from the emulsion picture, there was only one visible track coming out of the end of the pion track. Considering the *law of conservation of momentum*, there had to be at least one other particle born at the same time. This is because the pion, when it comes to rest, has zero momentum while the resulting decay muon clearly does have momentum. Since the total momentum must remain zero, this muon's momentum must be cancelled out by one or more invisible particles moving in opposite directions. This is like a gun's recoil in the opposite direction when a bullet is fired. These invisible particles must not carry any charge or they would leave a track in the emulsion. Using conservation of energy and conservation of momentum laws scientists concluded that the mass of the invisible particle was likely zero. There were only two known neutral particles of zero mass, the photon and the neutrino.

Neutrinos were extremely difficult to detect while photons can be detected easily by the secondary electrons they generate in matter, either by the Compton Effect or pair production. Experiments failed to find photons in the decay parts of pions. So, by the process of elimination it was concluded that the invisible particle was a neutrino (ν). The decay process for positively or negatively charged pions (π) to muons (μ) is described symbolically as follows:

$$\pi^+ \rightarrow \mu^+ + \nu \quad \text{or} \quad \pi^- \rightarrow \mu^- + \overline{\nu}$$

Experimenters used a similar process to find the decay products of muons by employing *artificial muons* produced in high energy accelerators. Leighton, Anderson and Seriff at the California Institute of Technology in 1949 showed that the muon's decay electrons were distributed over a very wide range. This indicated there were at least two invisible neutral particles in addition to the electron (or positron) in the decay particles. Further experiments showed that these particles must be neutrinos (ν). The decay process for muons (μ) to electrons (e) is described symbolically as follows:

$$\mu^+ \rightarrow e^+ + \nu + \overline{\nu} \quad \text{or} \quad \mu^- \rightarrow e^- + \nu + \overline{\nu}$$

It is interesting to note that since the charged pions decay into two particles, a muon and a neutrino or antineutrino, then conservation of momentum and energy give the decay products definite energies. This contrasts with the three-particle decay of the muon in which the emitted particles have a range of energies and momenta. Such detailed particle decay analysis was relatively new at this time. Today, the mathematical derivations of these classic pion decays are routinely assigned in physics classes as homework problems.

The pion story would not be complete without mentioning the *neutral* pion. This particle was postulated ever since the discovery of charged pions. The probable existence of a neutral companion to the charged Yukawa particle had been suggested by Nicholas Kemmer in 1938 on the grounds that nuclear forces should be independent of nuclear charge. He predicted that it should decay into two gamma rays with a lifetime of about 10^{-16} seconds. In 1947-48 it was suggested by Taketani in Japan, and by Rossi and Oppenheimer in the USA, that the decay of such a meson very rapidly into photons might be responsible for most of the cosmic ray photons and electrons found in the atmosphere.

If they existed, neutral pions would be more difficult to detect and observe than charged pions. Neutral pions would not leave tracks in photographic emulsions, or in cloud chambers. Despite that, the neutral pion was observed through its decay products (photons) - at the end of 1949 both in cosmic rays in Bristol and from an accelerator experiment by Steinberger, Panofsky and Steller at Berkeley. These experimental studies gave the neutral pion a value of 295 electron masses.

The decay process for neutral pions (π^0) to gamma rays (γ) is described symbolically as follows:

$$\pi^0 \rightarrow \gamma + \gamma$$

The mean life of a neutral pion was found to be 2 x 10^{-16} second. This is millions of times shorter than that of a charged pion. Because of the extremely short decay time, the two photons from the decay appear to originate exactly from the same point in the emulsion. Although photon paths are invisible, their paths can be derived from the pair production produced in the emulsion. It now became clear that the neutral pion plays an important role in starting electron-photon cascades in the atmosphere, since it decays almost immediately into two gamma rays. See Section 6.2 for more information on electromagnetic cascades.

Thus, with the pion identified, another piece of the cosmic ray puzzle was completed. At this point enough details of particle production were known to connect the primary radiation to the radiation normally found in the atmosphere. In the following sections the overall picture of this connection will be discussed

7.4 Nuclear Active Particles

In the early work on cosmic rays, scientists believed that cosmic radiation from space was responsible for the secondary radiation in the atmosphere. Many of them thought that these secondary particles were the result of collisions of the original particles with atomic nuclei. They hypothesized that these collisions would break up the nuclei with many negative and positive electrons being liberated. But such an explanation for particle showers did not fit the facts as known at the time.

It was observed that electrons and positrons appeared in single pairs not large groups of many pairs. It was known that the large number of particles in the showers were the result of many separate events involving radiation by electrons and pair production by photons. (See Section 6.4 for more information on cascade showers). Such occurrences happened in the vicinity of nuclei which provided the electric fields necessary for the processes but otherwise left the nuclei intact.

In spite of this some scientists found evidence for cosmic ray particles breaking up atomic nuclei. They believed that this process would give rise to groups of secondary particles completely different from electromagnetic cascade showers. The first indisputable evidence that the disintegration of heavy nuclei was real was the cosmic ray star discovered by Blau and Wambacher in 1932 using photographic emulsions. (See Section 7.2 for more information on this star). Of particular importance was that the stars were found more frequently at higher elevations.

For years these stars were the only direct evidence of the supposed interactions between cosmic rays and atmospheric atoms. Since these stars were rare it was thought that they did not play any significant role in the overall picture of cosmic rays. With the discovery of the pion the thinking changed dramatically. Because pions had such a short life they could not have come from any great distance. Thus, they must have been created in the atmosphere and were not part of the original radiation. So, how were they produced?

One possibility was that photons might produce pions in a way comparable to the materialization process of electron pairs. But theorists argued that the observed pions were more likely the result of nuclear interactions. It was suggested that pions and muons are produced in small groups and should be observed as showers of penetrating particles.

Some observers had seen penetrating showers on cloud chamber photographs, notably J. C. Street's student Lewis Fussell at Harvard in 1938. In Fussell's experiment, cosmic rays crossed the chamber in which three lead plates had been inserted. The thickness of the plates was chosen to enhance the probability that in each plate a simple process would occur; either the emission of a photon or the creation of an electron positron pair. A few of Fussell's pictures, about 3 out of the 900 taken, showed showers diverging from a well-defined point that were unaccountable by cascade theory. These pictures were discussed at a cosmic ray symposium in Chicago where some attendees attributed them to nuclear bursts.

In 1939, J. C. Street constructed another multiplate cloud chamber for the same purpose of observing ordinary cascade showers of photons and electrons. Among the large number of showers observed there were a few events of a different type in which cosmic rays apparently collided with atomic nuclei. A distinctive feature of these events was the birth of secondary particles that passed through the plates *without* producing electron showers. A typical meson (muon) shower is shown in Figure 7.10

Figure 7.10 This is a classic penetrating shower traversing several lead plates. A particle is seen entering the chamber from above and has a nuclear encounter in the second plate. It is probably a proton. Products of the interaction are four penetrating particles and also a weak cascade component which is completely absorbed in less than 6.5 cm of lead. Of the four muons, the one on the left stops in the chamber and the other three continue.

Around 1940 Gleb Wataghin, a Ukrainian scientist working in Brazil and Lajos Jánossy, a Hungarian physicist now working in England became seriously interested in this problem. Wataghin used a Brazilian Air Force plane to perform experiments at high altitude.

Wataghin's experiments used four GM counters located in the corners of a rectangle with the bottom two counters surrounded by lead as shown in Figure 7.11. He looked for four-fold coincidences in the counters, which occurred at a rate of one coincidence every five hours. Because of the dimensions of the setup and the arrangement of the counters he was able to show that these observations cannot be accounted for by either cascade showers or knock-on showers. The required energy to create such a penetrating cascade was estimated to be in the BeV range. This experiment showed conclusively that electrons were not responsible for the penetrating cascade.

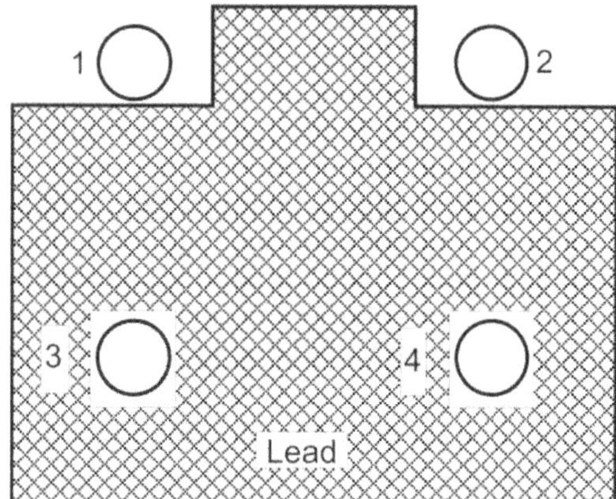

Figure 7.11 This is a typical arrangement used by Wataghin in 1940 to check for penetrating particles. Four-fold coincidences between the four counters cannot be due to cascade showers

Jánossy's experimental arrangement was more complicated than Wataghin's and consisted of a number of GM counters with electronic circuits to detect coincidences out of line. His counters were arranged so that a thick layer of lead could be placed between them. As many as five-fold coincidences were recorded frequently. Because of his unique arrangement of counters and lead, he was able to show that such coincidences cannot be accounted for in terms of either cascade showers or knock-on showers. He showed that a penetrating cascade reaching the bottom counters through 30 cm of lead required primary particle energies in the high BeV range. As noted by Jánossy, the possibility of electrons with this energy is much too small for the observed rate of one in five hours.

The detectors designed by Wataghin and Jánossy became known as *penetrating shower detectors* since they only respond to interactions initiated by particles of very high energies – many billions of eV. They proved to be very useful in the analysis of nuclear reactions caused by cosmic rays.

The nuclear emulsion method discussed in Section 7.2 was of critical importance in studying these nuclear interactions. When Kodak introduced emulsions sensitive to minimum ionizing particles it became possible to observe cosmic ray stars produced by high energy pions, see Figure 7.12. These pictures showed that when a very high energy particle collides with a nucleus it produced groups of pions and muons as the theory predicted. The details of the interaction at the collision point would have been extremely difficult to see with a multiplate cloud chamber due to the thickness of the metal plates.

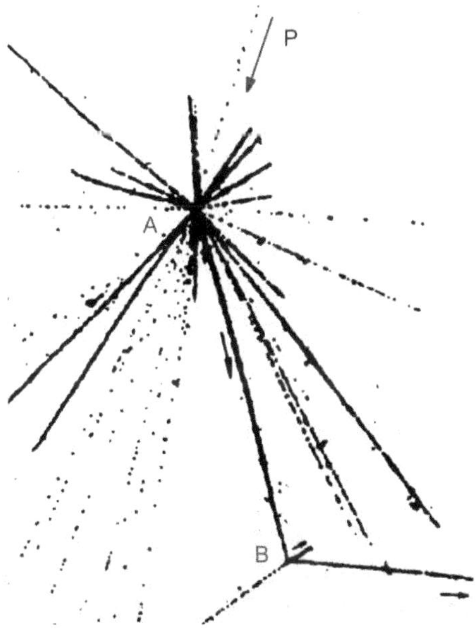

Figure 7.12 Typical cosmic ray star from a collision of high energy extraterrestrial particle P in an emulsion at high altitude. Collision occurs at point A and creates many additional particles. One of these causes another collision at point B.

So now we are back to the original question of what kinds of particles in the cosmic radiation are responsible for the nuclear interactions that were observed in these experiments. Studying the rate of occurrences of nuclear interactions with various absorbers and at different heights provided more clues.

In 1947 D. H. Perkins found that the number of cosmic ray stars was pretty much independent of any lead screen around a nuclear emulsion. The screen could be made thick enough to filter out most electrons and photons but yet produced stars.

John Tinlot, a student of Bruno Rossi at MIT, in 1948 used a penetrating shower detector to make a careful survey of altitude variations in the rate of occurrence of nuclear interactions. The instrument used by Tinlot was a detector of the type employed by Wataghin and Jánossy. His work was carried out at sea level and at various locations in Colorado aboard a B29 airplane. Figure 7.13 shows Tinlot's results for coincidences with respect to atmospheric height.

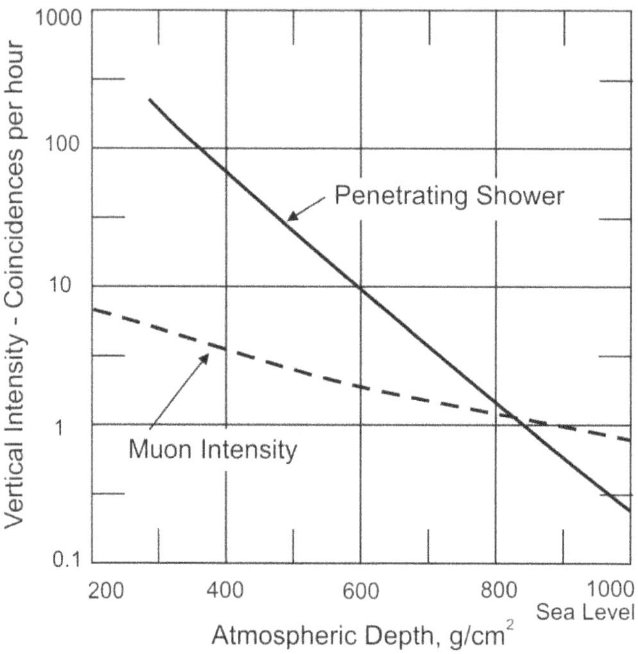

Figure 7.13 Comparison between the altitude dependence of penetrating showers measured by Tinlot and altitude dependence of muon intensity in 1948.

These results show that the number of high energy particles causing these penetrating showers increased with height faster than the number of muons. For example, from sea level to the top of Mt. Evans, at a height of 14,300 feet the muon intensity increases by a factor of 2.5 while the production of the showers increased by a factor of 32. Clearly muons could not be responsible for the penetrating showers.

All of the above results demonstrated two things. First, they showed that the bulk of the sea level radiation consisting of high energy electrons, photons and muons do not interact appreciably with atomic nuclei.

Secondly, they showed that in addition to the particles just mentioned, there must exist one or more different particles with the power to break up nuclei and produce even more particles. Although these *nuclear active particles* are sparse at sea level they increase rapidly with altitude.

Why are these findings important? They showed that many thousands of cosmic ray electrons, photons and muons can travel through the atmosphere without producing a single nuclear interaction. With respect to these particles an atom behaves like a cloud being penetrated by a bullet. Even the high energy cosmic ray photons and electrons do not interact with the nucleus. And, as was seen in Section 6.6, high energy muons are unable to interact with nuclei and are not easily absorbed by nuclei even when stopped in matter. On the other hand nuclear active particles are absorbed very fast in the atmosphere.

Scientists were very interested in determining the nature of the nuclear active particles. There were inklings that protons, neutrons or pions might be the particles in question. Since protons and neutrons interact very strongly in a nucleus they would likely interact with any nucleus they encountered while travelling through the atmosphere. Pions too play a strong role in the nucleus since it was found that they were produced in nuclear collisions.

Before leaving this subject it should be mentioned that scientists continued carrying out various new experiments on nuclear active particles. In the following years, cosmic rays scientists began detecting many more new, unstable particles. Initially there was confusion on how to deal with the number and properties of these new particles. But over the years the Standard Model was established with12 fundamental particles: six quarks and six leptons. There are currently hundreds of identified particles made from combinations of these twelve fundamental particles and scientists are still finding more.

Now let's see how nuclear active particles fit into the cosmic ray picture. This part of the story takes us to the top of the atmosphere and the primary radiation.

7.5 Radiation from Space

Near the end of 1950, with the discovery of pions in the atmosphere, a reliable and cohesive picture of all cosmic ray phenomena began to emerge. Scientists now had answers concerning the composition and interaction of the primary radiation that rains upon the Earth's atmosphere from outer space. Many researchers had a role in putting together this comprehensive picture of cosmic rays. Before we present the details of this picture, let's briefly review some of the work that led up to this point.

In the very beginning only a few scientists were concerned with determining the source of the mysterious ionization that was responsible for discharging a high quality electroscope. But things changed in the early 1930's when the German physicist Erich Regener perfected the technique of automatic recording electroscopes carried by balloons. These highly accurate measurements of cosmic ray ionization versus height finally convinced most scientists that the cosmic radiation was of extraterrestrial origin, see Section 2.3.

One of Regener's collaborators, Georg Pfotzer produced a confirming result in 1935 by sending up a sophisticated GM telescope. Unlike electroscopes which measure cosmic rays from all directions, his instrument pointed in the vertical direction and measured only a small area of the sky. His results are shown in Figure 7.14. Looking downward from the top of the atmosphere his curve shows a remarkable buildup of coincidences to a maximum at a depth of 90 g/cm^2. This can only be explained as an increase of secondary rays through interaction with primary rays. From the peak at 90 g/cm^2 downward, there is a gradual reduction in the coincidence rate to sea level.

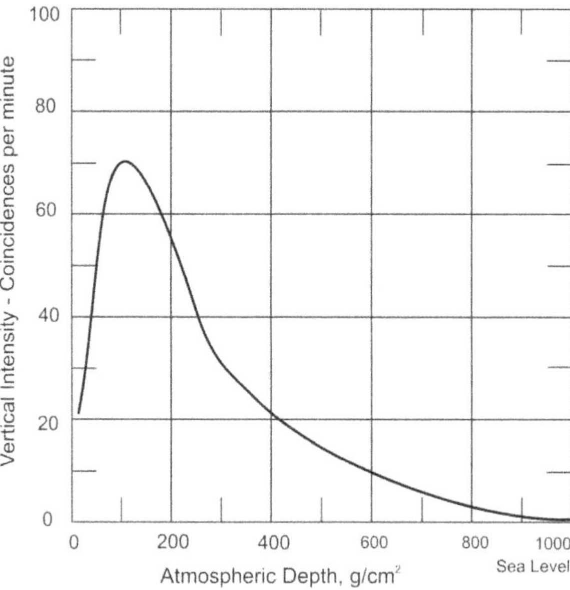

Figure 7.14 Vertical GM counter coincidence rate of cosmic rays depending upon atmospheric depth due to Pfotzer in 1935. The maximum, at around 90 g/cm^2, is now known as the Pfotzer peak.

These primary cosmic rays were *assumed* to be high-energy gamma rays (photons) mainly due to Milliken's influence, Section 4.0. But many physicists began to doubt this gamma ray assumption based on results from cloud chambers and Geiger counter experiments. Perhaps the most important Geiger counter experiment was that of Bothe and Kolhörster which established a case for the primary radiation being charged particles. These conflicting hypotheses about the primary radiation led to some heated debates.

Some scientists believed that if the primary cosmic rays were indeed charged, the Earth's magnetic field might influence them in some predictable way. On the other hand, if they are not charged particles, for example photons, they will proceed through the magnetic field without deviation.

As a consequence, in the 1930's the possible effect of the Earth's magnetic field on extraterrestrial cosmic rays was studied in great detail. Between 1931 and 1934, A. H. Compton and others organized heroic world surveys of cosmic rays which left no doubt that there was a latitude effect on cosmic rays due to the Earth's magnetic field. The existence of this effect implied that the primary cosmic rays were electrically charged. Soon thereafter, an east-west effect was discovered which proved that the primary rays were positively charged.

Some physicists erroneously identified the penetrating particles found near sea level as the primary radiation. However, by 1940 these penetrating particles were known to be muons which only lived for a few microseconds. Hence the muons could not be part of the radiation from outer space but had to be produced in the atmosphere by some other type of primary cosmic ray particle. Now, let's pick up the story from here.

Considering the known positive particles at the time, the primary radiation had to consist of electrons, or protons, or heavier nuclei, or a combination of all three. Although many scientists believed this to be true, until 1940 there was no experimental evidence to support such a conclusion.

Then, in 1941 just before the war, Marcel Schein and his collaborators at the University of Chicago reported the results of an especially significant experiment carried out on a balloon flight from Chicago. Their flight reached all the way up to 70,000 feet or 96 per cent of the atmosphere. They used GM tubes and lead absorbers and found that the high-energy particles passed through lead absorbers without generating the showers of low-energy electrons one would expect from a high-energy electron or positron. This finding - often overlooked in the literature due to the disturbed conditions prevailing in and after 1941 - demonstrated that the bulk of cosmic rays were probably protons.

Experiments on nuclear interactions of cosmic rays performed in the late 1940's endorsed this result as seen in Section 7.4. It was found that the number of *nuclear active* particles, such as protons, neutrons and pions, which constituted a small fraction of the total radiation found at sea level, increased with increasing altitude more rapidly than the number of electrons, photons and muons.

More experimental data obtained under various absorbers and at different altitudes convinced scientists that practically all primary cosmic rays were nuclear active particles. Of the nuclear active particles mentioned, only the proton was stable. Thus it was concluded by most scientists that *primary* cosmic rays were protons from space. For the most part this was true but as we'll see, some other particles needed to be included.

An important development along these lines occurred when two groups of cosmic ray scientists from the University of Minnesota (Edward Ney) and the University of Rochester (Hans Brandt and Bernard Peters) worked together in 1948 to study cosmic rays at high altitudes. They sent up sounding balloons carrying nuclear emulsions and a cloud chamber to 94,000 feet. Development of the plates revealed a number of very dense tracks, many of which had travelled through the entire stack of plates. Using emulsion plate analysis, the energy loss and total energy spent by the particles passing through the plates was calculated. It was found to be on the order of 1000 MeV and in some cases higher.

This presented a bit of a quandary. At such large energies, protons behave as minimum ionizing particles. But this was 2000 times the value predicted for such protons. Hence the observed particles were either heavier than protons or carried more charge. Further studies revealed that both the mass and the charge were larger than that of a proton. This indicated that the observed particles were the nuclei of elements heavier than hydrogen but stripped of their electrons. This was confirmed by the tracks of the small cloud chamber that was also taking photographs.

At this point most scientists were convinced that the primary cosmic radiation consisted of particles from space, namely protons and other heavy nuclei. These particles were not the result of secondary interactions in the atmosphere but belonged to the primary radiation itself.

The consensus today is that almost 90% of the cosmic rays that strike the Earth's atmosphere are protons and about 9% are alpha particles. Electrons amount to about 1% and there is a small fraction of heavier particles. About 0.25% of the cosmic rays are light elements, lithium, beryllium and boron. Lastly, a very small fraction of the cosmic rays are stable particles of antimatter, such as positrons or antiprotons. The precise nature of this remaining fraction is an area of active research.

In the next section some key questions relating to the cosmic radiation from space are discussed. What are the energies of these primary particles? What is produced in the top of the atmosphere from a collision? Answers to these important questions are part of the final pieces of the cosmic ray puzzle.

7.6 Completing the Cosmic Ray Picture

In the 1950's the cosmic ray puzzle was nearly complete. Scientists were now convinced that the primary cosmic radiation consisted of protons, other heavy nuclei and a fraction of other particles from space. The particles were found to be distributed over remarkably large energy values and a great a range of flux.

As these primary particles approach Earth they first encounter the geomagnetic field and their trajectories can be altered depending on where on Earth the particles arrive. The magnetic field is weakest at the poles allowing particles with all energies to enter the atmosphere at that point. From the poles to the equator the magnetic field increases preventing particles of increasing energy from reaching Earth. This is the so called latitude effect, Section 5.4. Deflection of the positively charged particles by the magnetic field also causes an asymmetry in the arrival directions producing more from the west than from the east. This is the so called east-west effect, Section 5.5.

Figure 7.15 shows the energy spectrum of primary cosmic ray particles as was known in the late 1950's. The horizontal axis shows the energy E of the cosmic ray particles in units of electron volts, eV. The vertical axis corresponds to the directional flux.

Notice the extremely large energy range portrayed for this curve, from 10^9 to 10^{19} eV. The portion of the curve between 10^9 and 10^{11} eV was obtained by balloon measurements based on the Earth's geomagnetic effect. Nuclear emulsions in balloons provided data between 10^{11} and 10^{13} eV. Data beyond 10^{15} eV is estimated from experiments on air showers. There are now research projects concentrating their efforts on detecting ultra-high energy particles beyond 10^{19} eV.

There is also a huge range of particle flux which varies from about 10^3 particles per square meter per sec per solid angle at 10^9 eV, to only 10^{-15} particles at 10^{19} eV, a factor of 10^{18} *smaller*. For instance, at about 10^{18} eV we can expect to receive one particle per square meter per solid angle *per year*. So, ultra-high energy particles are very rare.

In 1962 an ultra-high energy particle at the astonishing energy of more than 10^{19} eV was detected by John Linsley and Livio Scarsi in New Mexico as part of the MIT Volcano Ranch experiment. They used a giant array of sensors with an area of 2.2 km. Looking for higher energy particles such as this is still an active research area.

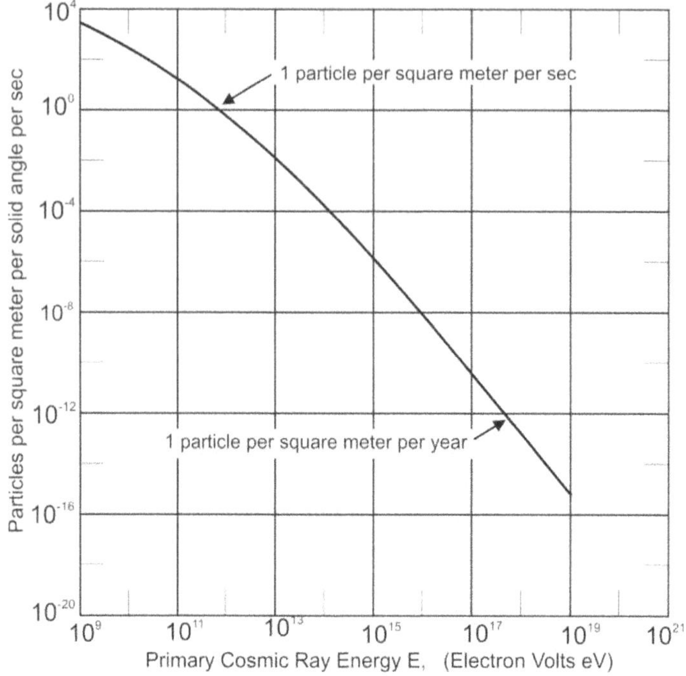

Figure 7.15 Energy spectrum of primary cosmic rays as known in the 1950's.

Particles entering the atmosphere soon collide with the nuclei of the atmospheric gases such as nitrogen and oxygen. At precisely what height in the atmosphere the collision occurs is a matter of probability. On average, protons collide after travelling about 70 g/cm^2 through the atmosphere. Given that the total atmospheric depth is about 1000 g/cm^2, this interaction depth corresponds to only about 7 per cent of the total air mass above sea level. Heavier particles like alpha particles collide sooner, travelling only an average 25 g/cm^2. Hence, given the large total atmospheric depth, the probability that a primary particle will reach sea level is effectively zero. Only at the highest mountain tops will one find a few surviving primary protons. Alpha particles and heavier nuclei are found only at the very top of the atmosphere and require balloon experiments for their observation.

Collision of the primary cosmic ray particle with a nucleus causes it to break up. Figure 7.16 shows the components and possible descendents of such a collision high in the atmosphere. There are *three modes* whereby the energy of the incoming particle is transferred through the atmosphere to sea level and below. They are the *nucleonic* component, the muonic or *hard* component and the electromagnetic or *soft* component. Depending on the magnitude of the primary particle, one of these three mechanisms provides for the conversion of primary energy into a secondary component.

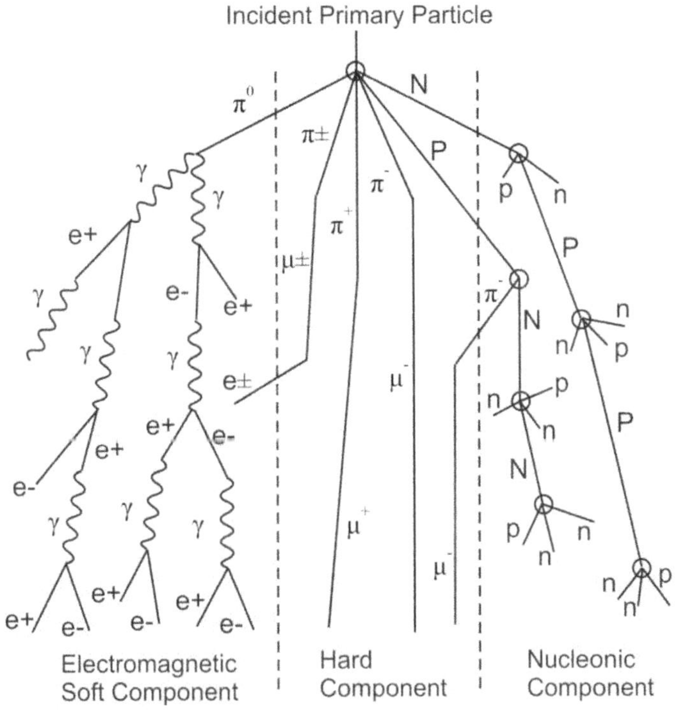

Figure 7.16 The descendants of an incident comic ray particle in the atmosphere. There are three major components, electromagnetic, muonic, and nucleonic. N and P correspond to high energy nucleons while n and p are disintegration products.

The *nucleonic* component is due to the interactions of lower energy primary particles with atoms in the atmosphere. This mechanism gives rise to neutrons and protons which are emitted as disintegration products of the interactions. This, in turn, produces a flux of particles which then develops into a cascade process. Upon reaching the lower atmosphere, the flux of this component decreases rapidly with atmospheric depth and comprises only a few percent of the particle flux at sea level. Nuclear interactions of these low energy primary particles result mostly in nucleons as disintegration products. A typical event in this sequence is the creation of a star as seen in Section 7.5.

At higher energies, a large portion of the energy of the primary particle goes into the creation of new, short lived pions, mesons and hyperons. For sufficiently high energies, pairs of protons and antiprotons as well as neutrons and antineutrons are produced.

Pions of all types are the most abundant of these secondary particles. Pions can be just as effective as protons and neutrons in producing nuclear interactions but *neutral* pions decay before they can produce such interactions. However, *charged* pions have a longer mean life and because they are moving close to the speed of light, where relativity theory applies, have a considerably extended lifetime. Hence, the probability that a charged pion will exist long enough to collide with a nucleus is higher and increases with the energy of the particle.

These secondary protons, neutrons, and pions together with other secondary particles and primary particles that survive, are responsible for the nuclear interactions at any level in the atmosphere. Nuclear active particles perish very rapidly with respect to atmospheric depth due to the small distances between collisions, rapid disappearance of pions due to decay and the energy loss at each interaction. Figure 7.17 shows the relative particle intensity for nuclear active particles, muons and high energy electrons, from the top of the atmosphere to sea level.

In the start of an electromagnetic shower, neutral pions decay quickly into photons. Then the photons disappear through the pair production process which materializes pairs of electrons and positrons. These electrons produce more photons. This is a cascade process that continually alternates between radiation and pair production as described in Section 6.4. The intensity of this process starts from zero at the top of the atmosphere, reaches a maximum at 120 g/cm^2 and then decreases rapidly as shown in Figure 7.17. This decrease in intensity occurs because the showers produced in the upper atmosphere gradually dissipate their energies in lower energy showers. Lower in the atmosphere there are no more initiating particles available to start new showers.

Muons that arise from decaying charged pions are free riders since they do not interact with atmospheric nuclei. They continue their journey losing energy only by ionization until they finally decay. Since the ionization loss in the atmosphere is small and their mean lifetime long, most of them reach sea level. This accounts for the slower decay of muons shown in Figure 7.17. Consequently, muons are the dominant component of the secondary comic radiation at sea level. Muons are very powerful and can bury themselves deep in the ground or lakes. Some muons decay in the atmosphere producing a few electrons found at sea level.

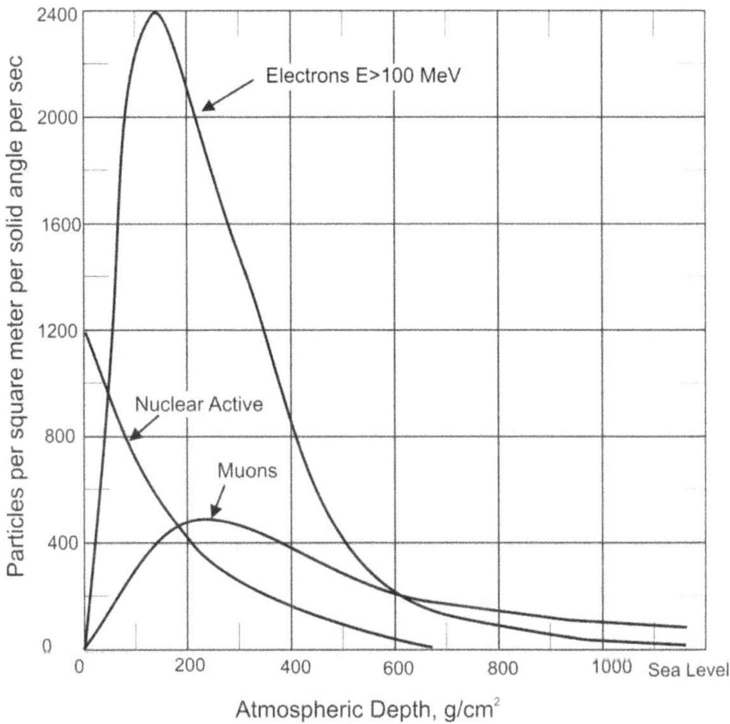

Figure 7.17 This is the vertical intensity of nuclear active particles, muons and high energy electrons plotted from the top of the atmosphere to sea level.

This completes the cosmic ray picture as it was understood in the late 1950's. The puzzle pieces have now fallen into place. Many thanks are due to all the scientists that have contributed to this amazing quest. We too are now witnesses to that wonderful journey and hopefully have obtained a deeper understanding of cosmic ray phenomenon.

It was way back in 1912 when Victor Hess and others put forward a hypothesis that there existed a penetrating radiation originating from outside the Earth's atmosphere. From this beginning, the field of cosmic ray physics has followed a winding but well defined path exploiting that original discovery for over fifty years.

During this period the study of cosmic rays evolved several major areas. One of these was the study of elementary particles and their interactions at high energy. Many of the unstable particles like the pion, muon and others were discovered as a result of cosmic ray experiments. These studies used traditional tools like cloud chambers, GM counters, nuclear emulsions and of course cosmic rays.

The beginning of a change was signaled with the arrival of artificially produced pions by the Berkeley cyclotron in 1948. By 1956 cosmic rays could no longer compete with accelerators for particle studies involving energies below 3000 MeV. Consequently, many workers found the uniqueness of cosmic rays in this area of research diminishing. Thus, many groups who were traditionally trained cosmic ray scientists became absorbed into these high energy physics laboratories.

At higher energies there remained a considerable interest in cosmic ray particles and their interactions. Energies of these rays up to the highest known values of 10^{19} eV were useful not only for the studies of nuclear cascades and extensive showers but for exploring the properties of these ultrahigh energy particles. The nature and origin of the primary particles became important to the field of astrophysics. Contributions of the early cosmic ray scientists led directly into the fields of x-ray, gamma ray and neutrino astronomy.

Cosmic rays are still an exciting field of study. Over the years many new and productive areas for the study and application of cosmic rays have emerged. Many of these interesting subjects are discussed later on in this book.

This concludes the first part of this book - the historical era of cosmic rays. We will continue now with a brief summary of the field and a number of other aspects of cosmic rays.

Chapter 8 to Chapter 15 The Modern Era

8 Cosmic Rays - From the 1950's to Today

8.1 The Cosmic Ray Field

Today, cosmic ray physics covers a wide range of disciplines ranging from particle physics to astrophysics and astronomy. Those considering entering this field should consult appropriate text books and teachers that are familiar with this subject.

The previous chapters described the early history of cosmic ray discovery and the research that led to the discovery of positrons, muons, pions, and other unusual particles. Electro-magnetic and nuclear-meson cosmic ray showers were shown to be a major part of generating these secondary components.

After World War II and into the 1950's the theory was firmed up and was now universally accepted. This physical understanding came from numerous experiments from scientists that were eager to explore this new energy regime. It was a completely new field – some say – the birth of particle physics.

These post-war years saw a steady increase in the number of research groups working in the cosmic ray field as well as the technical sophistication of the detectors. The early years had witnessed the development of experimental methods using diverse detectors for measuring air ionization, cloud chambers in a magnetic field, Geiger-Muller counters, photo emulsions and other schemes. But now the size of the cloud chamber dramatically increased with a typical arrangement of one chamber on top of the other in a high magnetic field and the bottom chamber consisting of a multiplate chamber to observe the resulting collisions or shower development. Large multiplate chambers were very popular in the USA.

Using two large superimposed Wilson chambers Clifford Charles Butler and G. D. Rochester (University of Manchester) found two unexpected events- one in October 1946 and one in May 1947- showing previously unknown particles. Their apparatus had a large magnetic chamber placed on top of a chamber fitted with copper plates. Nuclear reactions of cosmic rays occurred in a lead absorber placed immediately above the magnetic chamber, allowing short-lived secondary particles to be detected.

In order to increase the rate of detection with a higher flux of cosmic rays, the equipment was moved to the Pic du Midi de Bigorre observatory in the Pyrenees (Figure 8.1). In 1953, after the experiment had been running for a few months, the first two examples of K mesons passing through the first chamber and stopping in the second were observed. This was a first step in understanding of the quark structure of matter.

Figure 8.1 The double cloud-chamber, solenoid and control system in the Pic du Midi laboratory, as it was in 1953.

Emulsion techniques emerged after the war and were now simple and relatively easy to use (Figure 8.2). This led to new groups being formed in many countries, especially Italy. Technical advances allowed construction of large stacks of stripped emulsions, each one up to 600 um thick. These sheets, called pellicles, have no glass backing and can be stacked without spacing. Such stacks were easily flown in balloons. Afterwards the stack was disassembled and processed conventionally. A grid printed on the emulsion, using x-rays, made it possible to follow a particle track from one emulsion to another.

Collaboration grew between these new emulsion groups in the various countries, including several in Eastern Europe. They shared the cost of the large emulsion stacks as well as searching the emulsions and analyzing the events found.

Emulsion techniques were seen to be complimentary to the cloud chamber. Emulsions were particularly effective for the discovery and study of charged particles. Cloud chambers were better for neutral particles. Both methods were important in unraveling the complexities of the new particles being discovered

Figure 8.2 This nuclear emulsion stack of 240 sheets captures cosmic rays in three dimensions. These portable detectors are especially useful in a balloon or rocket launched cosmic ray experiment.

For two decades, cosmic ray studies gave us the first look into the world of subnuclear particles. However, things were changing with the advent of large particle accelerators in the 1950's. Consequently, cosmic ray studies slowly shifted into the background.

Particle accelerators have continued to become larger and more powerful. The latest accelerator, the Large Hadron Collider (LHC) of CERN, Geneva can attain energy in the 10 to15 TeV range. Ironically, this is still seven orders of magnitude *below* the largest cosmic rays.

High on mountain tops, aboard balloons, flying on high altitude aircraft, or included on space missions, cosmic ray experiments will continue to offer us a unique view of astrophysics at the highest level.

A lingering problem was determining the detailed nature of the primary cosmic radiation. Protons had been assumed the major component of primary cosmic rays. But now they could be studied with greater accuracy using rockets and spacecraft above the atmosphere. These missions provided a large boost to our knowledge because it was now possible to directly measure cosmic rays hitting the top of the atmosphere.

Just such an instrument was the high energy cosmic ray detector (Figure 8.3) designed at the Enrico Fermi Institute at the University of Chicago. It was designed to be flown in the Spacelab-Shuttle system in 1979. It was, at the time, the most sophisticated detector ever launched into space. It was designed to cover high energies up to 50 GeV and more. It used a combination of Cerenkov counters above and below the scintillators and a multiwire proportional chamber.

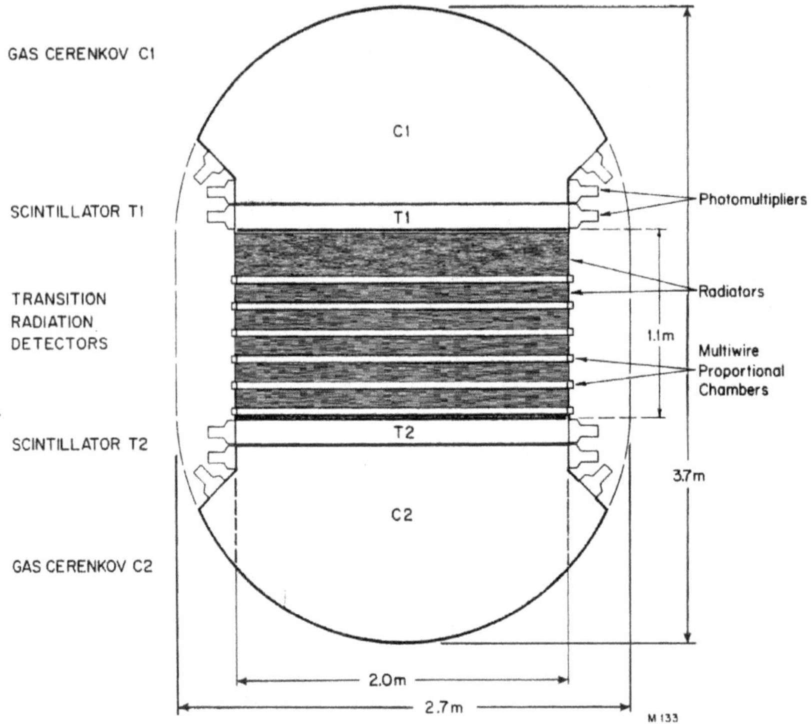

Figure 8.3 Cross section of high energy cosmic ray detector for Spacelab II. It was designed to be flown on the space shuttle in 1979.

However, detecting particles with very high energy in spacecraft has its own problems. Because the cosmic ray flux drops substantially with energy, several quite different types of detectors are necessary to study cosmic rays over the whole energy range. Small area detectors, (100 cm-squared) work well for cosmic rays in the GeV energy range where there are lots of particles available. But for energies in the 100 TeV range, where there are very few particles, a much larger area detector is needed which is prohibitive in cost for spacecraft. This problem has been attacked by using large area detectors on the Earth and measuring extended showers. This will be discussed in more detail later on.

8.2 Cosmic Rays Hitting Earth

Earth's atmosphere is hit by cosmic ray particles at the rate of approximately 1000 per square meter per second from outer space. These primary cosmic rays are ionized nuclei distinguished by their high energies and relativistic speeds. They come in many varieties and correspond to the familiar and stable subatomic particles that are normally found on Earth.

Cosmic rays have been found to contain mainly the nuclei of atoms like hydrogen, helium, carbon and so on but not complete atoms. This is because the electrons have been stripped off the atoms in the passage through space. In other words the cosmic particles consist of protons, nucleons consisting of protons plus neutrons, and a few other nucleons.

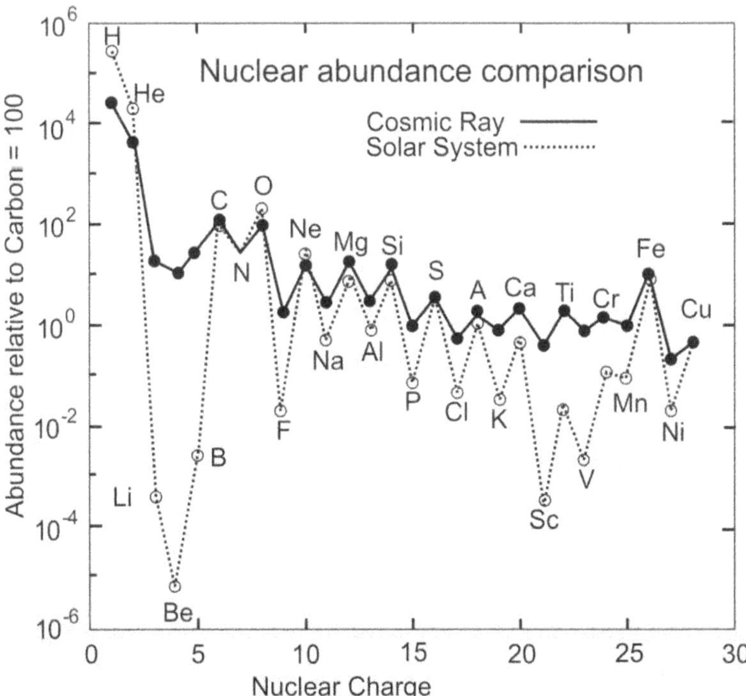

Figure 8.4 Abundance of nucleons in cosmic rays compared to those found in the solar system. Elements Li, Be and B show significant over-abundance.

All the nuclei of elements in the periodic table have been found in cosmic rays It has been found that 89% of the nuclei are hydrogen (protons), 10% helium (alpha particles: two protons and two neutrons), and about 1% other nucleons. Included is a very small fraction of other components such as electrons and a small amount of antimatter, such as positrons or antiprotons. The exact nature of this remaining fraction is a subject of current research.

In general, common heavier elements (such as carbon, oxygen, magnesium, silicon, and iron) are present in about the same relative abundances as in the solar system (Figure 8.4). Such is not the case for the light nuclei (lithium, beryllium and boron). These are very rare in the Universe. But when one looks at cosmic rays there is a huge over-abundance of these particles. The reason for this is still an active mystery in cosmic ray research.

8.3 Energies of Cosmic Rays

As we go to higher energies the number of cosmic rays hitting Earth declines dramatically as seen in the spectrum curve of Figure 8.5. The spectrum also exhibits a *knee* and an *ankle*, both of which deviate from the standard exponential decline curve (dashed line).

Notice that the graph does not use linear spacing, where each tic mark represents an equal amount. Instead it uses logarithmic (power of ten) spacing in order to compress all of the data into the plot. Otherwise the plot would be gigantic in size.

The range of energies encompassed by cosmic rays is truly enormous, starting at about 10^7 eV and reaching 10^{20} eV for the most energetic cosmic ray ever detected.

We see that the bulk of cosmic rays are in the lower energy range. For example at 10^9 eV (1 GeV) the particles arrive at Earth at about 1000 particles per square meter per second. At 10^{11} eV (100 GeV) there is about 1 particle per square meter per second. The curve continues to show a severe drop off of particle rate with energy.

At the ankle, corresponding to 10^{19} eV, the arrival rate is only about $(10^{-24)}$ particles per square meter per second. A 1 square kilometer detector would only count about 1 particle per km^2 – year. So these particles are rare but of enormous energy since 10^{19} eV equals = 10,000,000,000 GeV.

Cosmic Rays with energies above 10^{18} eV are referred to as ultra-high energy cosmic rays (UHECR). Interestingly, several particles in this energy range have been observed and will be discussed later on in this book.

Various mechanisms for producing changes in the steepness of the spectrum curve are the subject of intense study. Researchers assume that they are distinguished between populations of cosmic rays originating via different mechanisms.

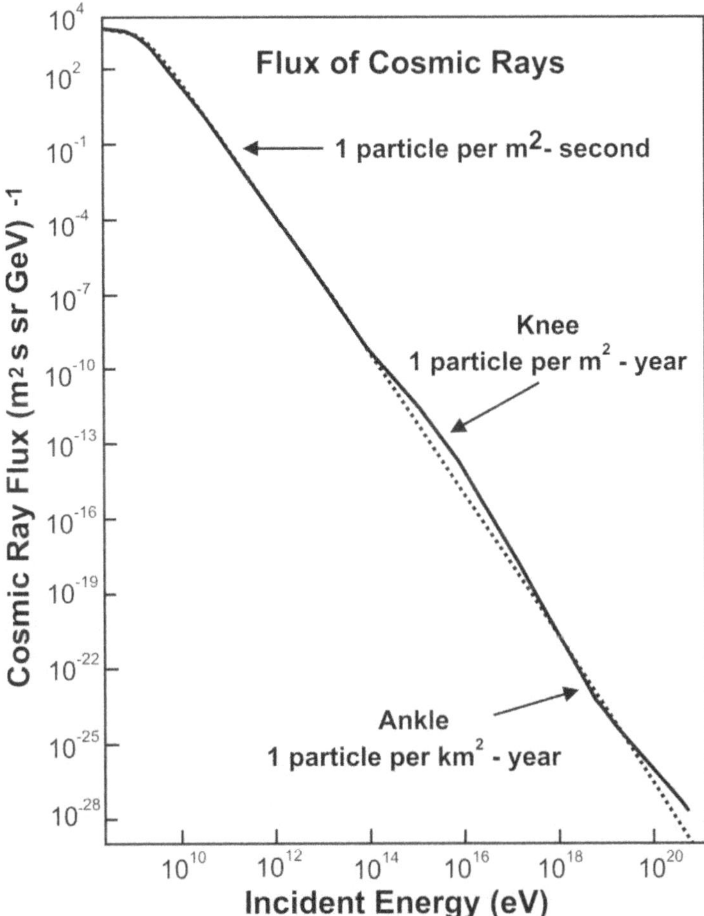

Figure 8.5 Cosmic ray energy spectrum of incident particles hitting Earth.

Current suggestions are that the cosmic rays with energies less than about 10^{10} eV are solar cosmic rays produced in solar flares and coronal mass ejections, while those with energies between 10^{10} eV and the *knee* at 10^{15} eV are galactic cosmic rays produced in the shocks of supernova remnants.

The origin of the cosmic rays with energies between the *knee* and *ankle* is unclear. They are again thought to be produced within the Galaxy, but the energies are too high for them to be accelerated by the shocks of supernova remnants. The origin for the ultra-high energy cosmic rays below the *ankle* is also a mystery, but many have suggested that these may be created outside of our Galaxy.

In the above discussion it seems like the particle rate is very low for the highest energy particles. But since the graph is based on a detection area of 1 square meter, we can obviously increase the number of particles detected by increasing the detection area. Doubling the area doubles the count rate and so on. However if we are trying to detect the particles from space with satellites, we cannot afford to orbit very large area detectors because of cost issues.

Hence in recent years there have been intense efforts to use ground based cosmic ray detectors. Recall that for ground based measurements we are looking at secondary cosmic rays generated in our atmosphere. From the ground measurements we need to infer the energy, composition, direction and other details of the primary cosmic particle. This is much more complicated and requires extensive modeling, but it can be done. Several ground-based cosmic ray projects discussed later on in this book use very large detection areas and operate in this manner.

8.4 Detecting Cosmic Rays

Cosmic rays are detected in a variety of ways depending on their spatial location and how much energy they possess. These methods may be classified as direct or indirect as discussed below. Generally speaking direct detection is more accurate than indirect detection. However, as we have seen in the previous section, the flux of cosmic rays decreases with energy. So we see that direct detection of high energy cosmic rays in space is hampered because of their small flux and the huge areal size required of the detectors. Thus, different techniques are used for direct and indirect methods. The appropriate regimes for each of these methods are shown in Figure 8.6.

Direct detection method. This method is associated with examining primary particles in outer space above or outside the influence of our atmosphere. It is generally used for particle energies in the range 10^3 to 10^{15} eV. High altitude balloons with instruments were first used for this purpose. Today direct detection is possible by all kinds of particle detectors at the International Space Station (ISS), on satellites, or high-altitude balloons. They are still hampered by the constraints in weight and size limiting the choices of detectors.

The first cosmic ray data ever transmitted from space was from Explorer 1, which launched in January 31, 1958, from the United States. It was the United States' first successful satellite launch. That data came from a cosmic ray detector on board the satellite that was invented by physicist James Van Allen (Figure 8.7).

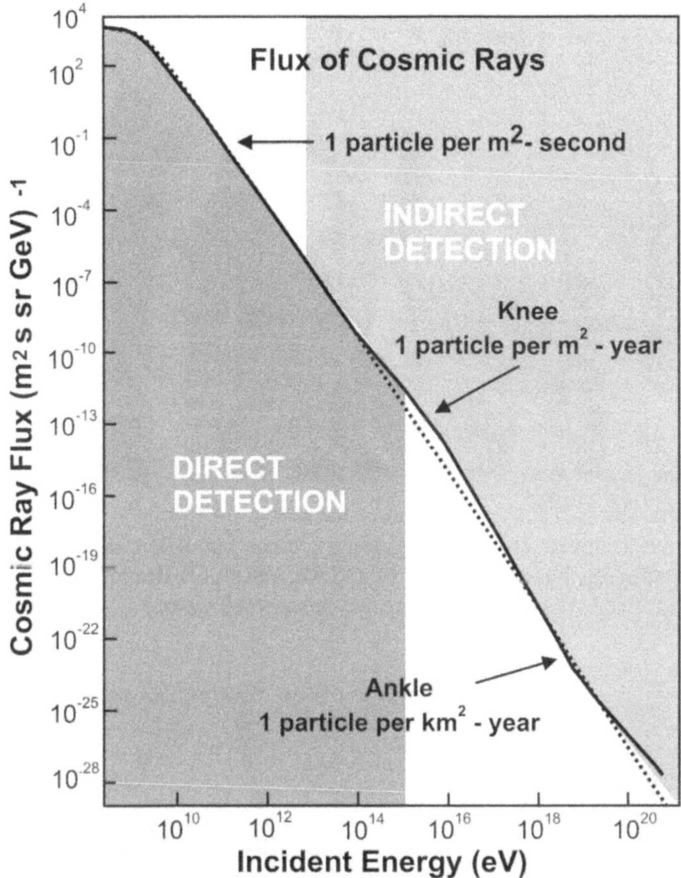

Figure 8.6 Regions of the energy spectrum most appropriate for utilization of direct or indirect cosmic ray detection methods. Notice that there is an overlap where both methods may be appropriate.

Explorer 1 was a huge success. It continued to operate for 12 years or 58,376 orbits. Explorer 1 carried instruments designed for the study of cosmic rays, micrometeorites, and for monitoring of the satellite's temperature (Figure 8.8). It was the first spacecraft to successfully detect the durably trapped radiation in the Earth's magnetosphere, dubbed the Van Allen Radiation Belt. Later satellite missions (in both the Explorer and Pioneer series) were to expand on the knowledge and extent of these zones of radiation, and were the foundation of modern magnetospheric studies.

Figure 8.7 From left: Bill Pickering, head of Pasadena, California's Jet Propulsion Laboratory which designed and built Explorer 1, James Van Allen, designer of the cosmic ray detector on board Explorer 1, and Wernher von Braun, whose team developed the Jupiter-C first stage (rocket) that launched Explorer 1.

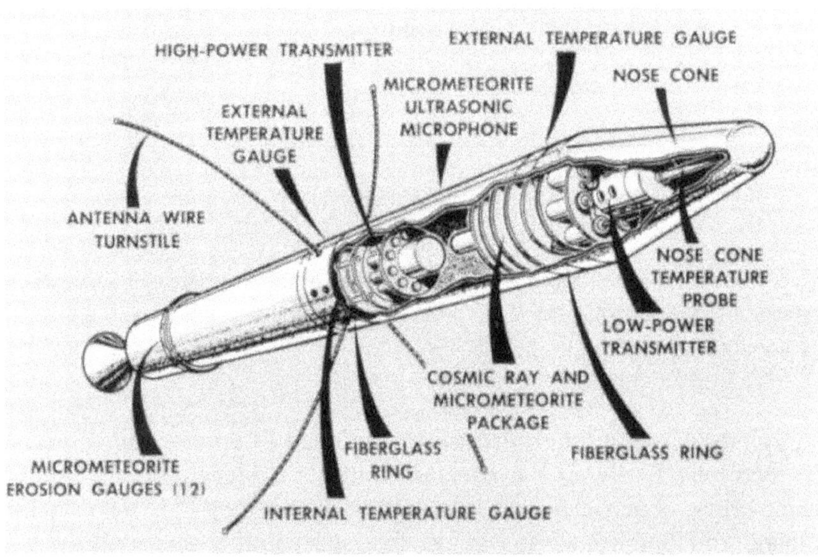

Figure 8.8 An artist's rendition of Explorer 1 and its components. Credit: NASA

One of the most interesting balloon programs was the "Cosmic Ray Energetics and Mass" (CREAM) experiment which began in 2004 by the University of Maryland and a team of collaborators (Figure 8.9). It was designed and constructed to measure cosmic ray elemental spectra using a series of flights. The goal was to extend direct measurement of cosmic-ray composition to the energies capable of generating gigantic air showers which have been mainly observed on the ground, thereby providing calibration for indirect measurements. These balloons were launched from McMurdo Station in Antarctica and were designed to stay aloft 60-100 days gathering data on charges and energies of the unimpeded cosmic rays that strike the detectors.

Figure 8.9 CREAM balloon launched from McMurdo Station in Antarctica.

A major advantage of this type of experiment is that it is possible to identify the original particle that would have caused the air shower detected by ground-based detectors. Accurate measurement of the composition of cosmic rays is necessary in order to understand the origins of the cosmic rays found above the "knee" at 10^{15} eV.

The CREAM mission has had seven successful flights from 2004 through 2016. Their 40 million cubic foot (1.1 million cubic meter) balloons carried each payload to its float altitude between ~38 and ~40 km.

ISS-CREAM is the next generation version of the CREAM balloon experiments that was sent to the International Space Station on August 14, 2017 with the CRS-12 mission, and will be installed permanently to the station. Located at an altitude of 410 km, 10 times higher than previous balloon flights, ISS-CREAM will be able to take data almost non-stop during its three-year mission. Because of the extreme altitude, there is no atmosphere for incident particles to scatter off of before reaching the detector. It is expected that this ISS-based mission will gather an order of magnitude more data than the CREAM balloon experiments.

NASA's vision for space exploration includes a return to the moon and beyond, with an eye for safe landing sites, potential resources, and characterizing the environment. A first step in this lunar endeavor was the launch of the Lunar Reconnaissance Orbiter (LRO) on June 18, 2009 (Figure 8.10).

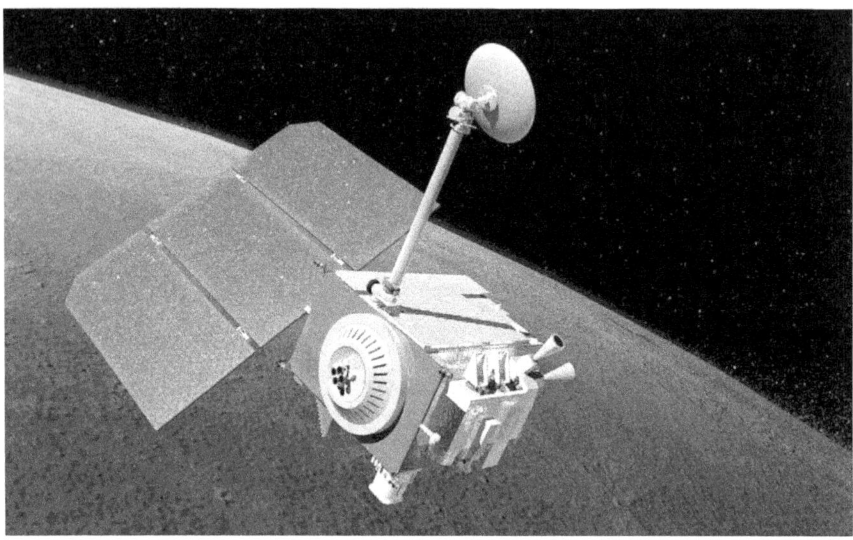

Figure 8.10 The Lunar Reconnaissance Orbiter (LRO) is a NASA robotic spacecraft currently orbiting the Moon in an eccentric polar mapping orbit. It is carrying a cosmic ray detector to characterize the radiation environment.

LRO's science payload of seven instruments gathers data useful for further exploration of the moon. Among those instruments is CRaTER Cosmic Ray Telescope for the Effects of Radiation. CRaTER characterizes the global lunar radiation environment and its biological impacts by measuring galactic and solar cosmic ray radiation behind a "human tissue-equivalent" plastic.

Twin *Voyager 1* and *Voyager 2* spacecraft are exploring where nothing from Earth has flown before (Figure 8.11). Continuing on their more-than 40-year journey since their 1977 launches, they each are much farther away from Earth and the sun than Pluto.

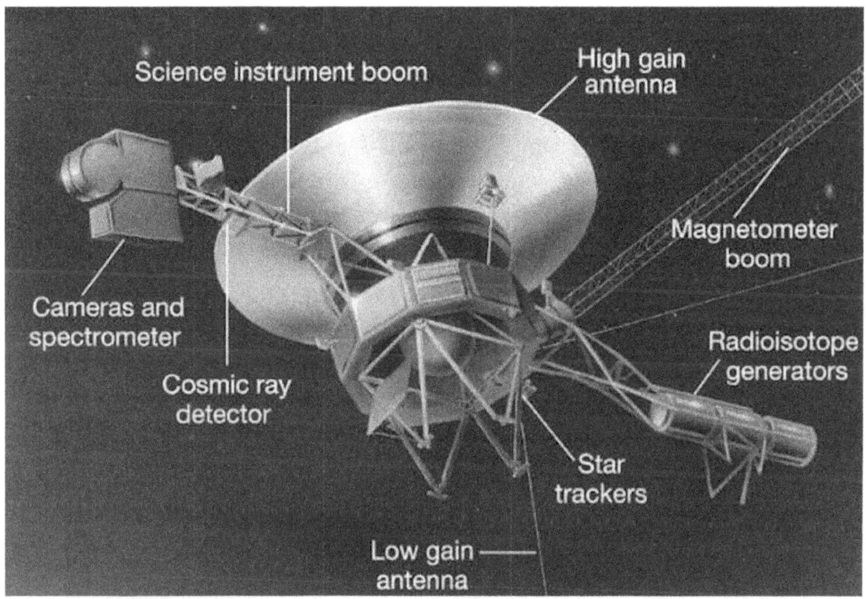

Figure 8.11 Voyager space probe launched by NASA on September 5, 1977 to study the outer Solar System. It is the most distant man-made object from Earth.

In August 2012, *Voyager 1* made the historic entry into interstellar space, the region between stars, filled with material ejected by the death of nearby stars millions of years ago. *Voyager 2* entered interstellar space on November 5, 2018 and scientists hope to learn more about this region. Both spacecraft are still sending scientific information about their surroundings through the Deep Space Network, or DSN as of April 24, 2021.

Aboard the *Voyager 1* and *Voyager 2* spacecraft of the NASA Voyager program is a Cosmic Ray System (CRS) to detect cosmic rays. The CRS includes a High-Energy Telescope System (HETS), Low-Energy Telescope System (LETS), and The Electron Telescope (TET). It is designed to detect energetic particles and some of the requirements were for the instrument to be reliable and to have enough charge resolution. It can also detect the energetic particles like protons from the Galaxy or Earth's Sun.

As of 2019, CRS is one of the active remaining instruments on both *Voyager* spacecraft, and it is described as being able to detect electrons from 3–110 MeV and cosmic ray nuclei 1–500 MeV/n. All three systems used solid-state detectors. CRS is one of the five fields and particle experiments on each spacecraft, and one of the goals is to gain a deeper understanding of the solar wind. Other objects of study include electrons and nuclei from planetary magnetospheres and from outside the solar system.

There are many other noteworthy direct measurements of cosmic rays by balloon and satellite such as PAMELA built by the Wizard collaboration, which includes Russia, Italy, Germany and Sweden. PAMELA was mounted on the upward-facing side of the Resurs-DK1 Russian satellite. It was launched by a Soyuz rocket from Baikonur Cosmodrome on 15 June 2006. This durable module remained operational and made significant scientific contributions until 2016.

Finally we must mention the Alpha Magnetic Spectrometer (AMS-02), a particle physics experiment module that is mounted on the International Space Station (ISS) (Figure 8.12). This experiment is a recognized CERN project that has a detector that measures antimatter in cosmic rays, crucial information needed to understand the formation of the Universe and search for evidence of dark matter.

Figure 8.12 Alpha Magnetic Spectrometer (AMS-02) mounted on the ISS.

A group led by MIT particle physicist Samuel Ting proposed this project in 1995. It was launched by the Space Shuttle *Endeavour* flight STS-134 carrying AMS-02 on 16 May 2011, and the spectrometer was installed on 19 May 2011. By April 15, 2015, AMS-02 had recorded over 60 billion cosmic ray events and 90 billion after five years of operation.

This development effort now involves the work of 500 scientists from 56 institutions and 16 countries organized under United States Department of Energy (DOE) sponsorship. It has been called "the most sophisticated particle detector ever sent into space", rivaling very large detectors used at major particle accelerators,

Indirect detection method. This method is associated with examining higher energy cosmic ray particles on Earth. It is generally used for particle energies in the energy range 10^{13} to 10^{22} eV and usually involves interaction of the particles in the atmosphere.

Higher energy cosmic rays are so rare that detection with sufficient statistics needs larger detection areas than direct measurements can provide. Hence indirect measurement methods based on the detection of extensive air showers induced by cosmic rays are used. These air showers consist of secondary particles generated in the atmosphere when a primary cosmic ray particle interacts with air molecules.

Air showers can be detected either by particle detectors on ground or by measuring electromagnetic radiation induced by the air shower particles in the atmosphere. The problem to be solved lies in the reconstruction of the primary cosmic ray particle properties such as its arrival direction, energy and mass.

A cosmic ray shower was first witnessed by Bruno Rossi through a unanticipated event in 1934 while testing his Geiger counters. He noticed "*that from time to time there arrived upon the equipment very extensive groups of particles which produce coincidences between counters even rather distant from each other*". Unfortunately because of commitments associated with his new position at the University of Padua he was unable to follow up on this discovery.

In 1938 Kolhörster and others extended the separation distance of counters for which coincidences were found to nearly 80m. But quite independently, and somewhat serendipitously, Pierre Auger and Ronald Maze also discovered the air-shower phenomenon in Paris in 1938. Their apparatus was of high precision in coincidence timing. With this equipment they went to Pic du Midi (2877m) and other mountains where they were able to demonstrate coincidences with counters 300m apart. This established beyond any doubt the occurrence of air showers and Auger was the first to provide some preliminary information about the primary particle responsible for the shower.

In 1948 Robert Williams of MIT made some observations at Mount Evans, Colorado using an array of four pulse type ionization chambers suggested by Bruno Rossi. He succeeded in determining the core location and the approximate lateral distribution of shower particles for many air showers. In one shower he found about 10^8 particles corresponding to a primary energy of 2×10^{17} eV.

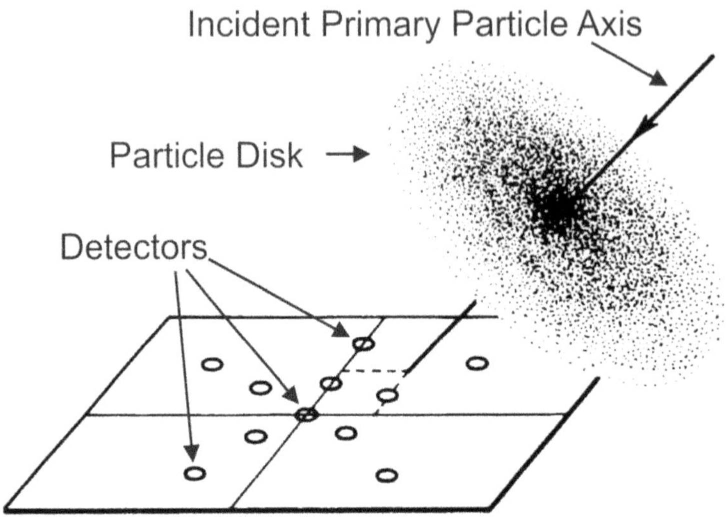

Figure 8.13 Shower disk approaching detectors (shown as circles). The energy and direction of the primary particle can be inferred from the detector data received.

In 1955 George Clark and his collaborators built the first air shower array at the Agassiz Station of the Harvard Observatory. A small air shower array is shown in Figure 8.13. Agassiz's array of 15 liquid scintillators was arranged in a ring of 500m diameter. Digital computers were used for the first time to analyze the particle data. The scientists detected a large shower of 2.6×10^9 particles corresponding to primary energy of 5×10^{18} eV.

In 1959 a team led by John Linsley of MIT constructed the first giant array. It was located in a remote area of New Mexico at Volcano Ranch. This project consisted of 19 detectors, each with an area of 3.3 square meters, spread over a ground area of 8 square kilometers. In a 6-month period about 10,000 showers of more than 10 million particles was observed. Among them were two of nearly 10 billion particles each. The largest shower detected at Volcano Ranch had about 5×10^{10} particles, corresponding to a primary energy of about 10^{20} eV.

Since this time a number of ground array experiments designed to measure cosmic rays in the ultra-high energy range have been carried out. These include Haverah Park array in England, Yakutsk array in Russia, the Sydney University Giant Airshower Recorder (SUGAR) in Australia, KASCADE experiment in Germany and its follow-up KASCADE-Grande, and the Akeno Giant Air-Shower Array (AGASA) in Japan.

Whereas many systems are designed to detect cosmic rays at ground level, there exists a class of detectors, loosely called "Fly's Eye", that use an array of mirrors and sensitive photo tubes to collect the faint light produced by the particles on their way down through the atmosphere.

A team from the University of Utah, led by George Cassiday began observations in 1981 with a new "Fly's Eye" telescope at Dugway Proving Ground, 160 km southwest of Salt Lake City (Figure 8.14). The team constructed 67 mirror units each with a cluster of 12 to 14 photomultiplier tubes in the mirror focal plane.

Figure 8.14 Utah Fly's Eye cosmic ray mirror telescope units and buildings.

It was built to record the passage of ultra-high energy cosmic rays through the atmosphere via atmospheric fluorescence. Fly's Eye operated successfully from 1981 to 1993. Half-way through that period the team added a second eye to the system. Fly's Eye II was constructed about 3 km from Little Granite Mountain on the valley floor. It consisted of 36 mirror units covering only half of the night sky, but in the same direction as the original (Figure 8.15).

On 15 October 1991 the Fly's Eye cosmic-ray detector in Utah, USA, recorded the highest-energy cosmic ray ever detected. The researchers measured the energy of the unusual cosmic ray event to be 3.2×10^{20} eV.

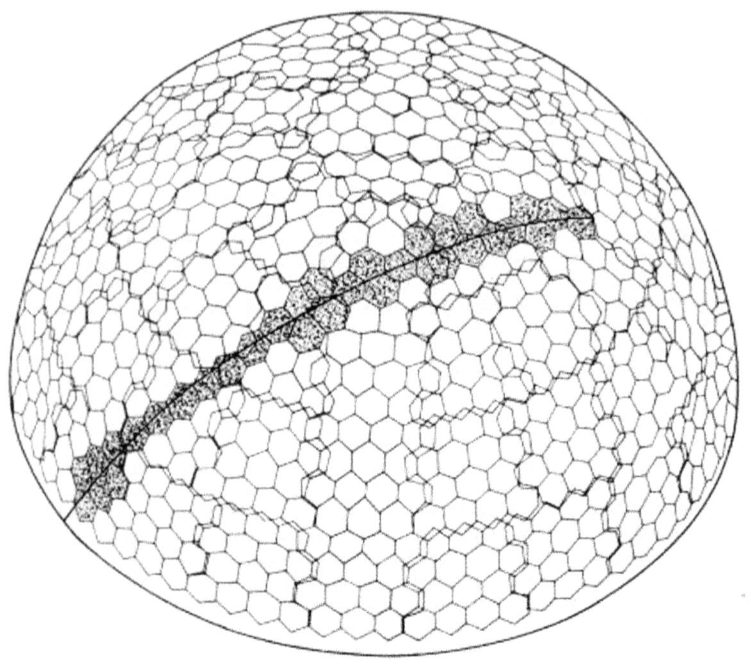

Fig 8.15 Utah Fly's Eye detector collected images of the night sky with mirrors and photomultiplier tubes. This diagram shows each hexagonal pixel representing a tube. An air shower is detected when a series of pixels sees light as shown here by the filled pixels producing the streak from left to right.

In 1999 the immense Pierre Auger Observatory was begun in Argentina by a group of 300 under the guidance of Jim Cronin (University of Chicago). This huge project was conceived by Jim Cronin and Alan Watson (Leeds University) in 1991 to address the mysteries of the origin and nature of the highest-energy cosmic rays. They reasoned that only a very large detector would have the exposure to collect enough events to answer these questions.

The observatory was named after Pierre Auger and today more than 500 scientists from nearly 100 institutions around the world are collaborating to maintain the site in Argentina as well as collect and analyze the data.

The Pierre Auger Observatory records cosmic ray showers through an array of 1,600 particle detectors placed 1.5 kilometers (about one mile) apart in a grid spread across 3,000 square kilometers (1,200 square miles). There are also twenty-four specially designed telescopes to record the emission of fluorescence light from the air shower. The combination of particle detectors and fluorescence telescopes provides an exceptionally powerful instrument for this research.

Figure 8.16 This is one of the water tanks used as a Cherenkov detector for the Auger project. It was chosen because of robustness and low cost.

Water tank Cherenkov detectors are used to measure the energy flux of electrons, photons, and muons in the shower (Figure 8.16). Each of the 1600 surface detector stations includes a 3.6 m diameter water tank with 12,000 liters of pure water and a sealed liner with a reflective inner surface. Cherenkov light produced by the passage of particles through the water is collected by three nine-inch-diameter photomultiplier tubes

Air fluorescence telescopes, similar to the Fly's Eye, are used to detect weak light emitted isotropically from the shower moving through the atmosphere. These fluorescence detectors comprise 24 large telescopes specialized for measuring the nitrogen fluorescence caused by charged particles of cosmic ray air showers (Figure 8.17). On dark moonless nights, these telescopes record the development of what is essentially the electromagnetic shower that results from the interaction of the primary particle with the upper atmosphere.

Figure 8.17 Auger project fluorescence telescope building. Each of the 4 buildings houses 6 fluorescence telescopes that overlook the general area.

The Pierre Auger Observatory is unique as it is the first experiment that combines both ground detectors and fluorescence detectors at the same site thus allowing cross-calibration and reduction of systematic effects that may be peculiar to each technique.

8.5 Current Picture of Cosmic Rays

Now let's summarize the main features of cosmic radiation. Cosmic particles arrive at Earth from all directions in space and impact the air molecules at the top of the atmosphere. They arrive at a rate that depends on the latitude, due to the effect of the Earth's magnetic field. The arrival rate at the equator is about 1000 particles per square meter and increases to ten times this amount in the northern latitudes. Particle flux is nearly constant and does not vary by more than a few tenths of a percent, except during rare solar events.

The arriving particles are mostly charged having a composition and energy as described in Section 8.3, with most of them being protons. Lowest energy particles have the greatest flux with the flux decreasing rapidly with energy. As shown in the Figure 8.18, the order of magnitude of flux particles covers a very large span. The highest energy particle has an energy of 10^{20} eV or nearly 16 joules.

Cosmic ray particles do not travel in straight lines in space as they are bent by galactic and intergalactic magnetic fields. Once in the solar system they are bent further by magnetic fields of the Sun and the Earth.

Particle energy (eV)	**Particle rate ($m^{-2}s^{-1}$)**
1×10^{9} (GeV)	1×10^{4}
1×10^{12} (TeV)	1
1×10^{16} (10 PeV)	1×10^{-7} (a few times a year)
1×10^{20} (100 EeV)	1×10^{-15} (once a century)

Figure 8.18 Relative particle energies and rates of cosmic rays

This flux of incoming cosmic rays at the upper atmosphere is dependent on the solar wind, the Earth's magnetic field, and the energy of the cosmic rays. At distances of about 90 astronomical units (AU) the solar wind undergoes a transition, called the termination shock, changing from supersonic to subsonic speeds. This region acts as a barrier to cosmic rays, decreasing the flux at lower energies ($\leq$ 1 GeV) by about 90%. But the strength of the solar wind is not constant, so it is observed that cosmic ray flux is correlated to some extent with solar activity.

Earth's magnetic field allows only high energy particles to enter at the equator. However, at the Earth's poles, particles of low energy are permitted entry along the magnetic field lines. This magnetic deflection of cosmic rays from the surface gives rise to a latitude effect – an increase of intensity from the equator to the poles. Moreover the geomagnetic field deflects particles of opposite sign in different directions – called the east-west effect- with slightly more particles coming from the west. Spacecraft readings have confirmed that most of the primary radiation coming from the west is positive.

Primary particles entering the atmosphere interact with air nuclei (protons and neutrons) releasing many types of mesons in groups of many particles. These pions, kaons and other particles decay rapidly forming showers. Most of the primary cosmic rays are absorbed in the top-most tenth of the atmosphere.

Muons, a hard component of radiation at sea level, are formed in the upper atmosphere. Because of their high energy, relativistic dilation permits them to travel through the atmosphere undergoing low energy loss, due primarily to excitation and ionization. They are found in great numbers at sea level and even at great depth below ground.

Total intensity of the radiation increases with distance from the top of the atmosphere. It reaches a maximum of three to five times the primary intensity at a height that varies with the atmospheric pressure. This usually occurs at a pressure of one-tenth atmosphere. Once past this maximum the total intensity decreases progressively until at sea level it is only one-fiftieth of the maximum value.

From the maximum intensity point, the hard component and the soft component decline differently. At the maximum point, the hard component is only about one-fifth of the soft part. The hard component decreases continually from the maximum point following a near vertical direction and ends up with only about one-fiftieth of its maximum.

At sea level the total radiation consists of one-fourth soft and three-fourths hard component. Neutrons are also in the shower as well as the flux of neutrinos originating from the sun, stars or decay products from the decay of mesons and muons. For more details see Section 7.6 "Completing the Cosmic Ray Picture".

Chapter 9

9 Origin of Cosmic Rays

9.1 Introduction

Bruno Rossi, a prominent cosmic ray scientist, said in 1964, *"Half a century after the discovery of cosmic rays the problem of their origin is still unsolved"*. Well, more than another 50 years have gone by and we still do not know for certain where they come from and how they acquire their enormous energies. Understanding their origin remains one of the most important problems in astrophysics.

Over the years various models of the origin of cosmic rays have been presented. Thus far no model fully encompasses all of our knowledge. The long running competition between scientists on this subject offers a fascinating picture of how science progresses.

Models are vital because they can be used to explain the composition and energy distribution of the primary radiation. Interstellar cosmic rays are unique because these rays are a direct sample of matter from outside the solar system and contain elements that are much too rare to be seen in spectroscopic lines from other stars.

Most of our knowledge of the relative abundances of elements in the Universe is based on the optical spectra of stars and nebula, on the absorption of light in interstellar space and the chemical analysis of meteorites. But accurate evaluation of the relative abundances of different nuclei present in cosmic rays offers another path. This makes possible the study of the isotopic and elemental samples in the interstellar medium that would otherwise be unavailable.

Interstellar space contains complex fields that can distort the path of cosmic rays. As we study their means of propagation in interstellar space, we may learn more about the properties of the space.

Cosmic rays may travel over huge distances from anywhere in the Universe. The highest energy ones travel close to the speed of light. Figure 9.1 shows the distances to many well known celestial objects in light years. This gives us a good idea of the enormity of the observable Universe and possible regions for cosmic ray origination.

Today the 'supernova model' has the widest support. It provides a basis for planning future observations and evaluating the importance of each new piece of information. But before we discuss that, we'll take a look at some of the earlier theories for the origin of cosmic rays.

Figure 9.1 Distances travelled by light to various places in the Universe.

9.2 First Theories of Cosmic Ray Origins

It's easy to invent a hypothesis about the origin of cosmic rays but the difficulty is verifying whether or not it actually represents what occurs in nature. Ever since their discovery, speculation has been rife about how and where cosmic rays are born.

A great many ideas were proposed by many individuals contemplating sources from the sun to galaxies beyond our own and invoking acceleration mechanisms running the gamut from electric fields produced in stellar thunderstorms to the original act of creation of the Universe. Now let's take a look back at some of these early hypotheses.

In 1925 C. T. R. Wilson hypothesized that the origin of the high altitude radiation was due to the acceleration of beta particles by big electric fields inside thunderstorm clouds and between clouds and Earth. This theory was formulated when many scientists doubted the extra-terrestrial origin of the high altitude radiation. But, since Wilson was an expert in electric fields his theory was given some credence.

In 1928 Millikan and Cameron decided to test Wilson's theory at Lake Miguillais, which is surrounded in different directions by high mountains. Although the lake was protected from energetic ionizing particles, which could be formed in a thunderstorm, the measured ionization rate appeared the same as on a Californian beach during a thunderstorm. Thus Millikan and Cameron concluded that Wilson's theory was not valid.

In 1926 Millikan presented his view on the origin of cosmic radiation. He believed that cosmic rays are the birth cries of atoms being born in interstellar space, and that they present the first experimental evidence that the creator is still on the job. His theory was formulated when most scientists believed that cosmic rays were gamma rays. This theory was accepted for a few years but after it was shown that cosmic rays are charged particles it was abandoned.

There were even more theories. Bothe and Kolhörster believed that the particles were accelerated by the gravitational field in the Universe. Erich Regener thought they came from star's radiation while making their journey around the Universe. M. C. Holmes in 1937 proposed that ions created by sunlight could be separated by the Earth's electric field and accelerated. E. A. Milne put forth a complex theory in which cosmic rays are present since the very beginning being a part of a complex system that is accelerated by gravitational fields in cosmic space. All of these theories would eventually be abandoned.

Figure 9.2 Fritz Zwicky (left) and Walter Baade of Caltech in 1934 proposed that "bright novae" were a class of stellar explosion completely dissimilar from ordinary novas. These supernovas would occur much less often than novas, perhaps only a few times every 1,000 years in the Milky Way Galaxy. They correctly deduced that supernovae were the transitional form some stars took before becoming neutron stars, and that cosmic rays originated in supernovae. Their concept of neutron stars was so advanced it was not fully accepted until technological advances of the 1960's proved the hypotheses.

In 1934 Fritz Zwicky and Walter Baade, introduced the term supernova for short flaring, extremely bright astronomical objects. They noted that these objects emit visible radiation during the 25 days of maximum brightness with the total emission equivalent to 10^7 years of solar radiation. Their work demonstrated that supernovae may be good sources of cosmic rays, although they thought the cosmic rays were mostly positrons coming primarily from supernovas in other galaxies.

In retrospect it should be noted that the estimation by Zwicky and Baade of supernova energy going towards the generation of cosmic rays is overestimated by about two orders. In any case, this hypothesis did not find many followers at the time. Nevertheless, the idea of Zwicky and Baade of supernovae as the main sources of cosmic rays became very fruitful, and was accepted and developed later by many scientists.

A key problem of cosmic ray physics is where and how the particles acquire their immense energies. A major energy source must surely be needed to provide the high total energy density. Many scientists wondered if the sun or the stars were sources responsible for cosmic rays. This will be explored in the next sections.

9.3 Cosmic Rays from the Sun and Stars

We know that solar cosmic rays have energies of $\sim 10^7$ to 10^{10} eV and are ejected primarily in solar flares (Figure 9.3) and coronal mass ejections (CME). This acceleration occurs at times of high solar activity, when great eruptions take place and large masses of ionized gases shoot out from the sun into interplanetary space. These particles have a composition similar to that of the Sun, and are produced in the corona by shock acceleration, or when part of the solar magnetic field reconfigures itself.

An assumption one could make is that, since the sun produces cosmic rays, it naturally produces all cosmic rays. In other words it emits high energy particles continuously producing the normal cosmic ray flux observed on Earth. In this view, the solar flares are just temporary increases in its normal activity.

However, this argument fails since we see a uniform intensity of the radiation at all hours of the day and night. We see cosmic rays coming from all directions from everywhere in the sky.

It is true that the terrestrial magnetic field bends the trajectories of charged particles coming from the Sun allowing some of them to reach the night side of Earth. There are also weak magnetic fields throughout the solar system that can play a role. They can scatter charged particles randomly so they would not appear to come from the Sun even if they began there.

Nonetheless, scientists have found that if all cosmic ray particles came from the Sun neither the terrestrial magnetic field nor the interplanetary magnetic field could provide the almost perfect uniform flow that is observed. In addition, calculations on protons, with energies of 10^{15} eV or more, show they are not appreciably deflected by terrestrial or interplanetary magnetic fields. If they were produced by the Sun they would have to come in straight lines from the direction of the Sun. But air shower experiments have shown no preferred direction of arrival.

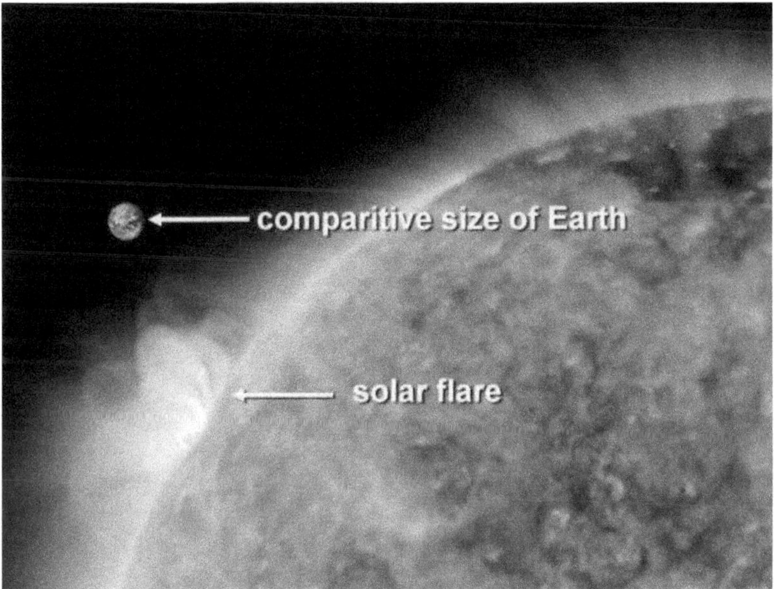

Figure 9.3 Solar flare on the Sun compared to size of Earth.

Since the Sun does not produce the major part of the observed cosmic radiation perhaps the stars could be the main source. Considering the billions of stars in the Universe, we could simply calculate the combined cosmic ray emissions of all these stars by assuming that the average star produces the same number of cosmic rays as the Sun. Considering the huge distance to the stars compared to the closeness of the Sun, calculations show that the total cosmic ray flux on Earth from the Sun is still millions of times greater than all the star's radiation put together. The situation is much like sunlight versus starlight intensity.

Of course all stars may not be like our Sun. There may be stars with special qualities that make them very effective cosmic ray sources. Possibly a relatively small number of these special stars could supply the whole observed cosmic ray flux.

Supernovas are very unusual stars and figure prominently as a cosmic ray source (Figure 9.4). These colossal explosions of individual stars occur only about once every 50 years in a galaxy the size of the Milky Way. Supernovae can briefly outshine entire galaxies and radiate more energy than our sun will in its entire lifetime. They're also the primary source of heavy elements in the universe.

Figure 9.4 SN 1987A was a type II supernova in the Large Magellanic Cloud, a dwarf satellite galaxy of the Milky Way. Supernova 1987A is clearly visible as the very bright star above the center.

It is now believed that galactic cosmic rays may be accelerated in shock waves initiated by supernova explosions. Other possible sources for cosmic rays include rotating neutron stars and black holes. Various ingenious theories have been presented to explain these ideas. We'll delve into these more later. But now let's review one of the earliest acceleration theories.

9.4 Fermi's Acceleration Theory for Cosmic Rays

Our Universe contains protons and heavier elements everywhere in space. The main problem faced by scientists in the early years was to figure out how they might have acquired their tremendous energies. At the time, commonly accepted acceleration mechanisms were solar explosions or remnants of supernovas.

In 1933, W. F. G. Swann suggested that the entire galaxy may be acting as a giant accelerator of cosmic rays. He suggested that cosmic radiation is most likely generated by acceleration of charged particles in electromagnetic fields in space or in stellar atmospheres. He pointed out that betatron acceleration may take place in the growing magnetic field of a sunspot. This is the same mechanism that was used later to develop the betatron – a type of particle accelerator.

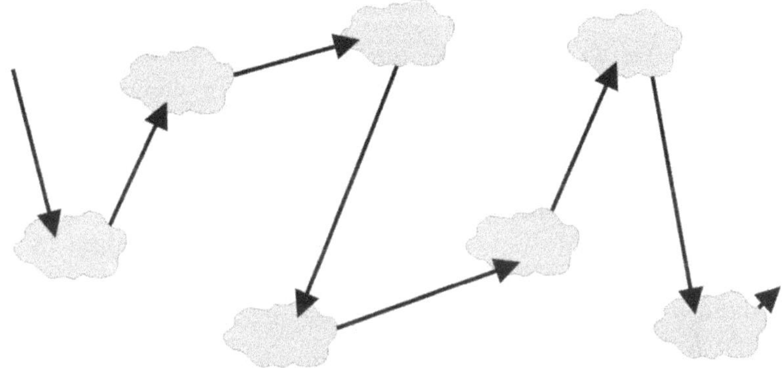

Figure 9.5 Simplified model of Fermi acceleration mechanism. It shows a cosmic ray particle transiting between moving magnetic clouds to gain energy.

In 1949 Enrico Fermi rejected the current idea that cosmic rays exist solely in our solar system and suggested a more realistic scenario based on the general picture of interstellar space that was emerging from astronomical observations. In this picture, enormous clouds of ionized gas, mostly hydrogen, were free to wander through space. These clouds contain magnetic fields, just like in solar flares ejected from the sun. Fermi showed that energy is exchanged between these wandering clouds and fast moving charged particles, and that on average the particles gain energy in the exchange.

To realize how this happens look at the simplified model shown in Figure 9.5. Assume that the clouds are rigid ionized gas bodies moving in space and supporting a magnetic field. The space between them is free of magnetic fields. Hence, a charged particle in this model will travel in a straight path until it enters a cloud. Inside the cloud it is subject to the magnetic field and will travel in a complicated path making a few or many orbits and finally emerge somewhere at a later time. Its original direction is now lost because of its complicated trajectory. Finally, according to the model, the particles repeat this action and scatter back and forth between clouds.

Fermi pointed out that when a particle collides with a cloud coming toward it; it effectively bounces off with increased energy. If it collides with a receding cloud it loses energy. And he showed that, on average, the particle obtains a small net gain, mainly because head-on collisions are more frequent.

Fermi calculated that the mean energy a proton would gain at each collision would be about 10 eV for the low energy ones and more for higher energy ones. In order for this to work, this gain must be larger than the ionization losses in the cloud. For protons, this will happen only if they have an initial energy of 200 MeV or more.

At the time this was very clever reasoning and it allowed scientists to derive the shape of the proton energy spectrum for comparison to the measured one. It also explained why it doesn't work for electrons since the ionization losses of electrons would exceed the gain.

As said, for this scheme to work it was necessary for protons to have energy of about 200 MeV. Fermi could not explain the origin of such energetic particles. The situation was even worse for heavier particles since the threshold was even higher.

A few years later it was found that the values in Fermi's model were not realistic based on astronomical data. In 1954 Fermi proposed a modified mechanism that considered the problem of injecting heavy nuclei.

Fermi's theory, while not being able to explain all the properties of cosmic rays, was still very useful to researchers in the field. Today Fermi acceleration is sometimes referred to as 'diffusive shock acceleration'. This is thought to be the primary mechanism by which particles gain non thermal energies in astrophysical shock waves. It plays a very important role in many astrophysical models, mainly of shocks including solar flares and supernova remnants.

9.5 Supernova Theory of Cosmic Rays

It is generally assumed that most cosmic rays come from sources within the Milky Way Galaxy. These are known as galactic cosmic rays (GCRs). The rest of the cosmic rays originate either from the Sun or, almost certainly, in the case of the particles with the highest energies, outside of our Galaxy. But why should we believe that is the case?

As we have noted previously there is a serious problem in identifying the source of cosmic rays. Being charged particles, their paths are curved both by magnetic fields in interstellar space and by Earth's own field. Consequently, cosmic ray sources cannot be identified from the direction of arrival but rather must be inferred by other means.

There are some clues about where cosmic rays might be generated. We know that magnetic fields in interstellar space are strong enough to keep all but the most energetic cosmic rays from escaping the Galaxy. It therefore seems likely that they are produced somewhere inside the Galaxy. The only likely exceptions are those with the very highest energy. Such cosmic rays move so rapidly that they are not significantly influenced by interstellar magnetic fields, and thus, they could escape our Galaxy.

This being the case, they could escape other galaxies as well, so some of the highest-energy cosmic rays that we detect may have been created in some distant galaxy. Still, most cosmic rays must have their source inside the Milky Way Galaxy.

Scientists have estimated how far typical cosmic rays travel before striking Earth. It is based on the light elements lithium, beryllium, and boron. These elements are formed when carbon, nitrogen, and oxygen strike interstellar protons - a reaction called spallation. They can calculate how long, on average, cosmic rays must travel through space in order to experience enough collisions to account for the amount of lithium and the other light elements that they contain. It turns out that the requisite distance is about 30 times around the Galaxy. At close to the speed of light, it takes perhaps 3 to 10 million years for the average cosmic ray to travel this distance. Since this is only a fraction of the age of the Galaxy (13.6 billion years), cosmic rays must have been created fairly recently on a cosmic timescale.

With an average life of 10 million years, GCRs must be replenished at an average power level of about 10^{41} ergs per second. Supernova explosions can supply this much power as they occur about every 50 years in the galaxy. While it appears that particle acceleration can be accomplished by the expanding shock waves from supernovas, details of the processes involved in cosmic ray production and acceleration remain unclear.

Some believe that charged particles can become trapped, bouncing back and forth across the front of the shock wave many times. With each pass through the shock, the magnetic fields inside it accelerate the particles more and more. Eventually, they are traveling at close to the speed of light and can escape from the shock to become cosmic rays. Some collapsed stars (including star remnants left over from supernova explosions) may, under the right circumstances, also serve as accelerators of particles.

Therefore many scientists believe that the best candidates for a source of cosmic rays are the Type II supernova explosions, which mark the violent deaths of some stars. A supernova happens when a star at least five times the mass of our sun goes out with a fantastic bang.

These massive stars burn huge amounts of nuclear fuel at their cores. This produces plenty of energy, so the core gets very hot. Heat generates pressure, and the pressure created by a star's nuclear burning keeps the star from collapsing.

Figure 9.6 A balance of gravity pushing in on the star and heat and pressure pushing outward from the star's core holds a star together.

In this condition the star is in balance between two opposite forces (Figure 9.6). The star's gravity tries to squeeze the star into the smallest, tightest core possible. But the nuclear fuel burning in the core creates strong outward pressure. This outward push resists the inward pull of gravity.

If you have a star with 8-25 times the mass of the sun, it can fuse heavier elements at its core. When it runs out of hydrogen, it switches to helium, and then carbon, neon, etc, all the way up the periodic table of elements. When it reaches iron, however, the fusion reaction takes more energy than it produces and it has run out of fuel.

When the star's fuel runs out, it cools off - causing the pressure to drop. Gravity wins out, and the star suddenly collapses. It is something like one million times the mass of Earth collapsing in a few seconds. It happens so quickly that it creates enormous shock waves that cause the outer part of the star to explode.

A very dense neutron star is usually left behind, along with an expanding cloud of hot gas called a nebula. A supernova of a star more than about 25 times the size of our sun may leave behind the densest object in the universe - a black hole.

Today, in spite of some unresolved problems, many scientists continue to favor the supernova source of cosmic ray generation in the galaxy. The evidence is strongly persuasive especially since there are no serious competitors.

9.6 Modern View of Cosmic Ray Origins

Today it is widely thought that cosmic rays originate from a wide variety of sources—ranging from energetic processes on the Sun all the way to the farthest reaches of the visible universe. Most cosmic rays originate from extra-solar sources within our own galaxy, such as rotating neutron stars, supernovae, and black holes.

However, some cosmic rays have extremely high energies, which provide evidence that at least some must be of extra-galactic origin possibly from radio galaxies and quasars. The local galactic magnetic field would not be able to contain particles with such a high energy. Cosmic rays with energies up to 10^{14} eV can be accounted for in terms of shock-wave acceleration in supernova shells. The origin of cosmic rays with energy greater than 10^{14} eV remains an active research area.

Cosmic rays with energy above 10 GeV (10 x 10^9 eV) approach the Earth's surface isotropically. It has been hypothesized that this is not due to an even distribution of cosmic ray sources, but instead is due to galactic magnetic fields causing cosmic rays to travel in spiral paths. Hence cosmic rays are not useful in positional astronomy, as they carry no information of their direction of origin. Below 10 GeV, there is a directional dependence, due to the interaction of the charged component of the cosmic rays with the Earth's magnetic field. Now let's summarize.

Solar Cosmic Rays. These are cosmic rays that originate from the Sun, with relatively low energy (10-100 keV per particle). The average composition is similar to that of the Sun. It has been found that solar cosmic rays vary widely in their intensity and spectrum, increasing in strength after some solar events, such as solar flares.

Moreover, an increase in the intensity of solar cosmic rays is followed by a decrease in all other cosmic rays, called the Forbush decrease, after their discoverer, the physicist Scott Forbush. These decreases are due to the solar wind, with its entrained magnetic field sweeping some of the galactic cosmic rays outward, away from the Sun and Earth. Forbush decreases tends to follow the 11-year sunspot cycle, but individual events are tied to events on the Sun.

There are differences between solar and galactic cosmic rays. This occurs mainly in the enhancement of heavy elements such as calcium, iron, and gallium, as well as cosmically rare light elements such as lithium and beryllium.

Galactic Cosmic Rays. These are high-energy charged particles that enter the Solar System from the galaxy. They are composed of protons, electrons, and fully ionized nuclei of light elements.

The bulk of galactic cosmic rays have energies too low to penetrate the Earth's atmosphere. Their trajectories in the Earth's magnetic field tend to channel them to the upper latitudes and the poles. These galactic cosmic rays are exactly like the charged particles that make up the solar wind. When they strike the atmosphere, they can create large showers of secondary particles, including exotic ones such as muons, and these secondary particles are what can be detected at the Earth's surface.

Some galactic cosmic rays have extremely high energy, leading to the inference that they must have originated in very energetic processes. Some are believed to have been accelerated by the shockwaves of supernovae. But some galactic cosmic rays have energies so high that it is difficult to visualize a physical process that could have created them.

Extragalactic Cosmic Rays. These are very-high-energy particles that flow into our Solar System from beyond our galaxy. The energies these particles possess are in excess of 10^{15} eV. Not much is known about the origins of extragalactic cosmic rays. This is partly due to the lack of particles. The number of particles reaching the Earth's surface originating from extragalactic sources is only about 1 particle per square meter per year.

Many theories abound as to the processes responsible for cosmic rays with such high energies. One theory is that cosmic rays gain more and more energy through electromagnetic processes. Bouncing randomly back and forth in the shock waves of some violent object allows some of the particles to gain energy. Eventually, they may build up enough speed to escape from the remnant. Possible sites considered include gamma ray bursts and active galactic nuclei.

Other possible sources considered by scientists for these energetic particles include colliding galaxy systems, accretion flow shocks to clusters of galaxies, and more exotic processes from the very early universe, such as the decay of superheavy particles trapped in the galactic halo, or topological defects.

Ultra-high-energy Cosmic Ray (UHECR). These are extraordinary cosmic rays that appear to have extreme kinetic energy, far beyond both its rest mass and energies typical of other cosmic rays. Typically it has energy greater than 10^{18} eV. These particles are significant because they have energy comparable to (and sometimes exceeding) the Greisen-Zatsepin-Kuzmin limit.

The GZK limit is a theoretical upper limit on the energy of cosmic ray protons traveling from other galaxies through the intergalactic medium to our galaxy. The limit is 5×10^{19} eV. (This corresponds to the energy of a proton travelling at ≈ 99.99999999999999999998 % the speed of light). The limit is set by slowing interactions of the protons with the microwave background radiation over long distances (≈160 million light-years).

Anomalous Cosmic Rays (ACRs). These are cosmic rays with unexpectedly low energies. They are thought to be created near the edge of Earth's solar system, in the heliosheath, the border region between the heliosphere and the interstellar medium. They are not discussed further in this book.

9.7 Ultra High Energy Gamma Rays and Cosmic Ray Origins

Ultra-high-energy cosmic gamma rays are of importance because they may reveal the source of cosmic rays. Ultra-high-energy gamma rays are gamma rays with photon energies higher than 100 TeV (100×10^{12} eV). The existence of these rays was confirmed in 2019. The highest energy astronomical gamma ray as of 2019 has its center in the Crab Nebula which is thought to contain a rapidly spinning neutron star, or pulsar (Figure 9.7).

Gamma rays travel in straight lines (like starlight) from their source to an observer. This is unlike cosmic rays which have their direction of travel scrambled by magnetic fields. Sources that produce cosmic rays will almost certainly produce gamma rays as well, since the cosmic ray particles interact with nuclei or electrons to produce photons or neutral pions which in turn decay to ultra-high-energy photons.

Figure 9.7 The Crab Nebula (designated M1, NGC 1952, Taurus A) is a supernova remnant and pulsar wind nebula in the constellation of Taurus.

Scientists tell us that the ratio of primary cosmic ray hadrons to gamma rays also gives a clue as to the origin of cosmic rays. While gamma rays could be produced near the source of cosmic rays, they could also be produced by interaction with the cosmic microwave background.

Ultra-high-energy gamma rays interact with magnetic fields to produce positron-electron pairs. In the Earth's magnetic field, a 10^{21} eV photon is expected to interact about 5000 km above the Earth's surface. These secondary electrons or positrons lose energy mainly by bremsstrahlung radiation to produce gamma rays. These produce more pairs and the cascading goes on until ionization becomes the main channel of energy loss.

Thus, upon the incident of each gamma ray on the atmosphere, a shower of charged particles is formed in the atmosphere. The air showers may be observed through the detection of light emitted (or induced) by the shower particles or the detection of those shower particles that manage to reach the ground.

Figure 9.8 The Tibet AS-gamma experiment detects high-energy gamma rays by observing showers of particles produced when a gamma ray hits Earth's atmosphere.

Located at an altitude of 4300 m above sea level in the town of Yangbajing in Tibet, the Tibet AS-gamma experiment has been operated jointly by China and Japan since 1990 (Figure 9.8). It involves 28 international institutions. Its research objectives include the measurement of the energy spectrum and chemical composition of very high-energy primary cosmic rays and the study of 3 dimensional global structures in the solar and interplanetary magnetic fields by means of high-energy galactic cosmic rays.

The air shower array consists of 697 scintillation counters which are placed in a lattice with 7.5 m spacing and 36 scintillation counters which are placed in a lattice with 15 m spacing. Each counter has a plate of plastic scintillator, 0.5 m^2 in area and 3 cm in thickness, equipped with a 2-inch diameter photomultiplier tube (PMT). The time and charge information of each PMT hit by an air shower event is recorded to determine its direction and energy. The detection threshold energy is approximately 3 TeV, which is the lowest one achieved by an air shower array in the world.

At the center of the air shower array, burst detectors and emulsion chambers were set up to closely observe the core region of an air shower event. Each burst detector is composed of a plate of plastic scintillator and 4 photodiodes attached to each corner of the plate. On each burst detector, are 6 layers of emulsion chambers (x-ray film interleaved with Pb plate). This hybrid experiment using the air shower array and burst detector emulsion chambers allows selection and measurement of the proton component in primary cosmic rays in the "knee" region (10^{15} to 10^{16} eV).

In 2021, the Tibet AS-gamma experiment discovered gamma rays beyond 100 TeV (10^{14} eV) from a supernova remnant (SNR), designated G106.3+2.7, 2600 light years from Earth.

These gamma rays are of the highest energy ever observed from SNRs, and are probably produced in collisions between cosmic rays (protons) accelerated in G106.3+2.7 and a nearby molecular cloud.

“SNR G106.3+2.7 is the first candidate object with sufficient evidence in the Milky Way that can accelerate cosmic rays (protons) up to 1 PeV (10^{15} eV),” said Huang Jing, one of the leading researchers of the study from the Institute of High Energy Physics (IHEP) of the Chinese Academy of Sciences.

Chapter 10

10 Large Cosmic Ray Experiments on Earth

10.1 Introduction

Over the years there have been a number of cosmic ray experiments on Earth. Some of them have finished their work but others are still operating at various places in the world. Several of them are discussed below.

10.2 Akeno Giant Air Shower Array (AGASA)

This is a very large surface array designed to study the origin of ultra-high-energy cosmic rays. Located in the town of Akeno in Yamanashi prefecture, Japan, it covers an area of 100 km^2 and consists of 111 scintillation detectors and 27 screened muon detectors. Array experiments such as this one are used to detect air shower particles. The array was operated by the Institute for Cosmic Ray Research, University of Tokyo at the Akeno Observatory. It has measured events from 1991 to 2004.

Results from AGASA were used to calculate the energy spectrum and anisotropy of cosmic rays. The results helped confirm the existence of ultra-high energy cosmic rays ($>5 \times 10^{19}$ eV), such as the so-called "Oh-My-God" particle that was observed by the Fly's Eye experiment run by the University of Utah, (Section 8.4).

The AGASA energy spectrum is shown in Figure 10.1 expanded in energy in order to emphasize details of the steeply falling spectrum. Numbers attached to points show the number of events in each energy bin. The dashed curve represents the spectrum expected for extragalactic sources distributed uniformly in the Universe.

This energy spectrum extends up to higher energies beyond previous results. Eleven events were observed above 10^{20}eV. Researchers believe that the energy spectrum is therefore more likely to extend beyond 10^{20}eV without the GZK cutoff. Their distribution of arrival direction and arrival time is quite uniform in view of the limited statistics.

On December 3, 1993 a very energetic cosmic ray of energy about 2×10^{20} eV was detected. This is the second highest energy particle traveling in the Universe which has ever been observed at this time. The particles in the air shower were spread over a 6km x 6km area. If this cosmic ray were a proton, its origin could be extragalactic!

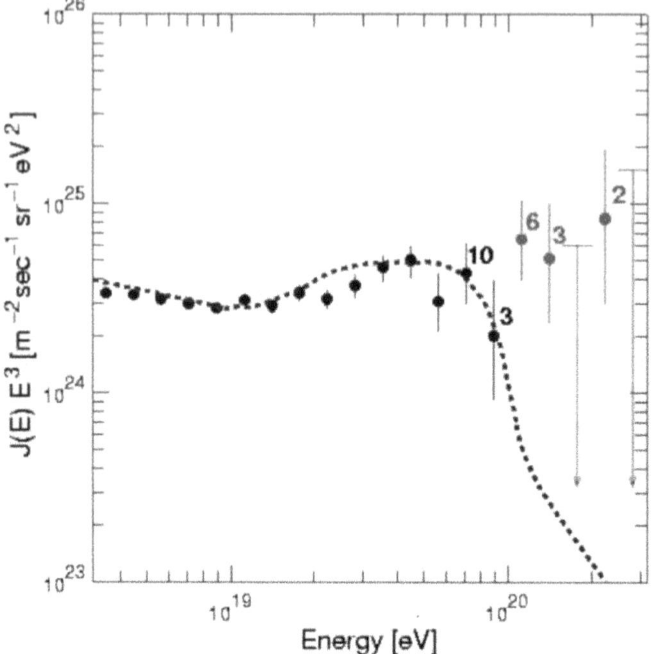

Figure 10.1 Energy spectrum determined by AGASA with zenith angles smaller than 45 degrees up until 2002. Error bars are 68% confidence limit. Numbers attached show the number of points in each energy bin. Dashed curve is spectrum expected for extragalactic sources distributed uniformly in the Universe.

AGASA and High Resolution Fly's Eye (HiRes) groups have merged to form the **Telescope Array** and with the Pierre Auger Observatory have improved on the results from AGASA by building larger, hybrid detectors and collecting greater quantities of more precise data. See later section.

10.3 Chicago Air Shower Array (CASA)

This was a significant ultra high high-energy astrophysics experiment operating in the 1990s. It consisted of a very large array of scintillation detectors located at Dugway Proving Grounds in Utah, USA, about 80 kilometers southwest of Salt Lake City (Figure 10.2).

The full CASA detector, consisting of 1089 detectors began operating in 1992 in conjunction with a second instrument, the Michigan Muon Array (MIA), under the name CASA-MIA. MIA was made of 2500 square meters of buried muon detectors. At the time of its operation, CASA-MIA was the most sensitive experiment built to date in the study of gamma ray and cosmic ray interactions at energies above 100 TeV (10^{14} eV).

Figure 10.2 Chicago air shower array of CASA-MIA experiment in Dugway Proving Grounds, Utah. The CASA scintillation detectors are the white square boxes laid out on 15-meter grid spacing. At the center of the array (left of center in image) is the Fly's Eye II detector. Photo courtesy of Kenneth Gibbs.

Research from this experiment covered searching for gamma rays from Galactic sources like the Crab Nebula, X-ray binaries, Cygnus X-3 and Hercules X-1. Other topics included searching for extragalactic sources (active Galactic nuclei and gamma-ray bursts), the study of diffuse gamma-ray emission (an isotropic component or from the Galactic plane), and measurements of the cosmic ray composition in the region from 100 to 100,000 TeV.

10.4 GAMMA Experiment

Construction of the GAMMA installation was begun in the middle of the 1980s on Mt. Aragats in Armenia under the direction of E. Mamidjanian (Yerevan Physics Institute) and S. Nikolsky (Moscow Lebedev Institute). It continued to grow and was updated through the years. In 2004 it was modernized again. It was producing data from 2003 – 2008 and was still operational and improving in 2010.

The ground-based GAMMA experiment is designed to study Extensive Air Showers (EAS) at 700 g/cm^2 of atmospheric depth in the primary energy range 10^2 - 10^5 TeV. It consists of an enlarged surface EAS array (108 scintillation detectors) and underground muon carpet (150 m^2 detectors) (Figure 10.3).

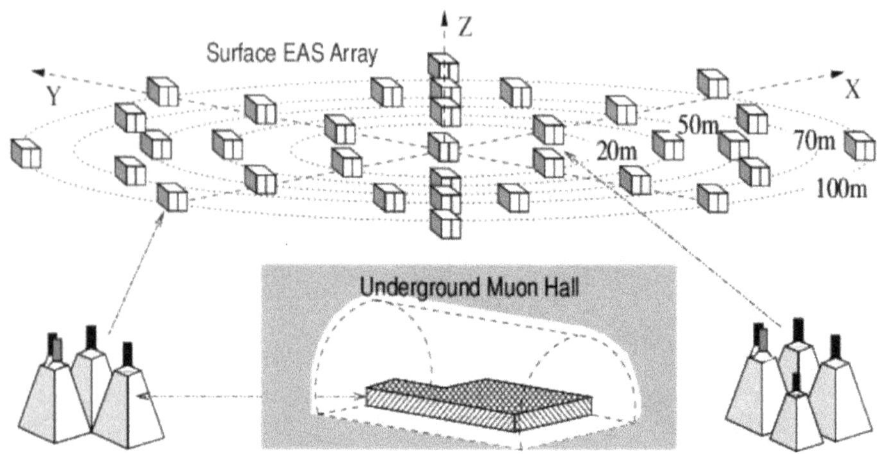

Figure 10.3 Layout of GAMMA experiment cosmic ray detectors.

Experiments are designed to study: (1) Primary cosmic ray energy spectra and elemental composition (abundances of the elements) at energies 10^{15} to 10^{18}eV (so called knee energy region), (2) Galactic diffuse gamma-ray intensity at energies 10^{14}–10^{15}eV, (3) Extensive Air Showers (EAS) at the mountain level by the ground-based EAS array and underground muon scintillation counters and, (4) Hard jets production at energies ~10^{16}eV by the muon multi-core shower events.

10.5 GRAPES-3 Experiment

This experiment (Gamma Ray Astronomy PeV EnergieS phase-3) located at Ooty in India started as a collaboration of the Indian Tata Institute of Fundamental Research and the Japanese Osaka City University, and now also includes the Japanese Nagoya Women's University.

Observations began with 217 plastic scintillators and a 560 m^2 area muon detector in 2000. Presently the array is operating with ~400 scintillators that are spread over an area of 25,000 m^2. The energy threshold of muon detectors is 1 GeV.

GRAPES-3 is designed to study cosmic rays with an array of air shower detectors and a large area muon detector (Figure 10.4). It aims to probe acceleration of cosmic rays in the following four astrophysical settings. These include acceleration of particles to: (1) ~100 MeV in atmospheric electric fields through muons, (2) ~10 GeV in the Solar System through muons, (3) ~1 PeV in our galaxy, (4) ~100 EeV in the nearby universe through measurement of diffuse gamma ray flux.

Figure 10.4 Air shower detector array of GRAPES-3 experiment at Ooty in India.

Presently GRAPES-3 is continuously operating with ~ 400 plastic scintillator detectors with a separation of 8 meters, to record the density and arrival time of particles in cosmic ray showers. At present, GRAPES-3 is the highest density conventional EAS array in the world and the associated huge 560 m^2 area tracking muon detector, is also the largest area tracking detector.

Results so far have been obtained on a variety of topics. They include: (1) measurement of primary composition in the energy range 50 TeV - 1 PeV, (2) precision measurements of Forbush decrease events, and (3) precision measurement of the density gradient of cosmic rays in the solar system.

10.6 High Altitude Water Cherenkov Experiment (HAWC)

This is a gamma ray and cosmic ray observatory located on the flanks of the Sierra Negra volcano in the Mexican state of Puebla at an altitude of 4100 meters (Figure 10.5). HAWC is the successor to the Milagro gamma-ray observatory in New Mexico, which was also a gamma-ray observatory based around the principle of detecting gamma-rays indirectly using the water Cherenkov method.

Figure 10.5. Each HAWC tank contains water and four photomultiplier tubes.

HAWC is a joint collaboration between a large number of American and Mexican universities. It is a wide view TeV gamma-ray telescope that explores the origin of cosmic rays and a variety of other studies.

HAWC detects electromagnetic radiation from air showers when high energy cosmic rays hit the Earth's atmosphere. It's sensitive to showers produced by primary cosmic rays with energies between 100 GeV and 50 TeV. It uses large metal tanks, containing a light-tight bladder holding 188,000 liters of water. Inside are four photomultiplier tubes (PMT) (3-8" and 1-10" high QE). High-energy particles striking the water result in Cherenkov light that is detected by the PMTs. HAWC uses the difference in arrival times of the light at different tanks to measure the direction of the primary particle. Light patterns allows for discrimination between primary (hadrons) and gamma-rays. From this, scientists can map the sky using gamma-rays.

10.7 High Energy Stereoscopic System (H.E.S.S.)

On 26 July 2012, the H.E.S.S. II telescope started operation in Namibia near Gamsberg, an area well known for its excellent optical quality. It is the largest Cherenkov telescope ever built, with five telescopes, four with a mirror just under 12 m in diameter, arranged 120 m apart from each other in a square, and one larger telescope with a 28 m mirror, constructed in the center of the array (Figure 10.6). It is a system for the investigation of cosmic gamma rays in the photon energy range of 0.03 to 100 TeV.

Figure 10.6. View of the full H.E.S.S. array with the four 12 m telescopes and the new 28 m H.E.S.S. II telescope standing above them all.

The instrument allows scientists to explore gamma-ray sources with intensities at a level of a few thousandths of the flux of the Crab nebula (the brightest steady source of gamma rays in the sky). The Cherenkov signal detected by photomultiplier tubes is sampled at 1 GHz.

In September 2018, the Collaboration released for the first time a small subset of its archival data. The release consists of event lists and instrument response functions for observations of various well-known gamma-ray sources (the Crab nebula, PKS 2155-304, MSH 15-52, RX J1713.7-3946) as well as observations of empty fields for background modeling.

10.8 High Resolution Fly's Eye Project

This project, also called the HiRes detector, was an ultra-high-energy cosmic ray observatory that operated in the western Utah desert from May 1997 until April 2006. It used the atmospheric fluorescence technique that was pioneered by the Utah group first in tests at the Volcano Ranch experiment and then with the original Fly's Eye experiment.

The HiRes experiment made the first observation of the (GZK cutoff) which is an indication of the highest energy cosmic rays interacting with the cosmic microwave background and the universe becoming opaque to their propagation. In 2010 final results of the HiRes experiment confirmed the GZK cutoff. See Chapter 8 for more details.

A follow-on experiment to the High Resolution Fly's Eye and Akeno Giant Air Shower Array (AGASA) experiments is the Telescope Array Project which began data collection in central Utah in 2007. A similar approach, but with the water-Cherenkov detectors, has been employed for the Pierre Auger Observatory which began collecting data in 2004.

10.9 Telescope Array Project

This project is an international collaboration involving research and educational institutions in Japan, The United States, Russia, South Korea, and Belgium (Figure 10.7). The experiment is designed to observe air showers induced by ultra-high-energy cosmic rays using a combination of ground array and air-fluorescence techniques. It is located in the high desert in Millard County, Utah, United States, at about 1,400 meters (4,600 ft) above sea level.

Figure 10.7 Telescope array project shower detector with the doors open.

The Telescope Array observatory is a hybrid detector system consisting of both an array of 507 scintillation surface detectors (SD) which measure the distribution of charged particles at the Earth's surface, and three fluorescence stations which observe the night sky above the SD array. Each fluorescence station is also accompanied by a LIDAR system for atmospheric monitoring.

The SD array is much like that of the AGASA group, but covers an area that is nine times larger. The hybrid setup of the Telescope Array project allows for simultaneous observation of both the longitudinal development and the lateral distribution of the air showers. When a cosmic ray passes through the Earth's atmosphere and triggers an air shower, the fluorescence telescopes measure the scintillation light generated as the shower passes through the gas of the atmosphere, while the array of scintillator surface detectors samples the footprint of the shower when it reaches the Earth's surface.

10.10 Pierre Auger Observatory

This is an international cosmic ray observatory in Argentina designed to detect ultra-high-energy cosmic rays: sub-atomic particles traveling nearly at the speed of light and each with energies beyond 10^{18} eV. In Earth's atmosphere such particles interact with air nuclei and produce various other particles. These air shower particles can be detected and measured.

But since these high energy particles have an estimated arrival rate of just 1 per km^2 per century, the Auger Observatory has created a detection area of 3,000 km^2 (1,200 sq mi) - in order to record a large number of these events. It is located in the western Mendoza Province, Argentina, near the Andes. See Chapter 8 for more details on this project.

10.11 IceCube Neutrino Observatory

This neutrino observatory at Amundsen-Scott South-Pole Station uses the clear, deep ice of the Antarctic glacier as a large target for detecting weakly interacting neutrinos, which are also expected from certain cosmic-ray sources, but which have so far not been found. The observatory is primarily designed to measure neutrinos from below, using the Earth as a filter to discriminate against muon background induced by cosmic rays

IceCube also includes an air shower array on the surface called IceTop extending IceCube's capabilities for cosmic ray physics (Fig 10.8). IceTop is an array of 81 stations spanning a square kilometer of the Antarctic ice sheet. Each station is located on top of one of IceCube's strings and holds two tanks of frozen water, each tank equipped with two IceCube sensors.

Construction of IceCube Neutrino Observatory was completed in December 2010. IceCube can be regarded as a cubic-kilometer scale three-dimensional cosmic ray detector with the air showers (mainly the electromagnetic component) measured by the surface detector IceTop and the high energy muons and neutrinos measured in the ice.

Figure 10.8 Aerial view of the IceTop detector near the Admunsen-Scott South Pole Station in Antarctica. Each circle shows an IceTop station as well as the location of an IceCube string.

In particular, the measurement of the electromagnetic component in IceTop in coincidence with the high energy muon bundle, originating from the first interactions in the atmosphere, has a strong sensitivity to composition. Here IceCube offers the unique possibility to clarify the cosmic ray composition and spectrum in the range between about 300 TeV and 1 EeV (10^{18}eV), including the 'knee' region and a possible transition from galactic to extra-galactic cosmic rays.

10.12 TAIGA Advanced Instrument Experiment

This project (formerly the Tunka experiment) measures air showers, which are initiated by charged cosmic rays or high energy gamma rays. It started in the 1990's with a small array of photomultiplier detectors and by 2020 has grown greatly in size. TAIGA is situated in Siberia in the Tunka valley close to Lake Baikal.

It consists of five different detector systems: Tunka-133, Tunka-Rex, and Tunka-Grande for charged cosmic rays; Tunka-HiSCORE and Tunka-IACT for gamma astronomy. From the measurements of each detector it is possible to reconstruct the arrival direction, energy and type of the cosmic rays, where the accuracy is enhanced by the combination of different detector systems.

The aim of the cosmic-ray measurements is to solve the question of the origin of the cosmic rays in the energy range up to about 1 EeV (10^{18}eV). For gamma-ray astronomy the aim is to identify sources of higher energy than possible by current gamma-ray observatories.

10.13 Large High Altitude Air Shower Observatory (LHAASO)

This is a gamma-ray and cosmic-ray observatory in Daocheng, in the Garzê Tibetan Autonomous Prefecture in Sichuan, China (Figure 10.9). It is designed to observe air showers triggered by gamma rays and cosmic rays. The observatory is at an altitude of 4,410 meters (14,470 ft) above sea level. Observations started in April 2019.

The observatory covers an area of 360 acres. It has three underground observing pools. Each of the pools contains 100,000 tons of water. The pools house 12 telescopes to capture high-energy photons. Cherenkov radiation detectors are used. Research teams from Australia and Thailand will participate in the project directly, with others expressing interest.

Figure 10.9 Aerial view of LHAASO in Sichuan, China.

The remarkable sensitivity of LHAASO in cosmic rays physics and gamma astronomy can play a key-role in the comprehensive general program to explore the High Energy Universe. As of 2021, LHAASO has found a dozen ultra-high-energy (UHE) cosmic accelerators within the Milky Way. It has also detected photons with energies exceeding 1 PeV (10^{15}eV).

10.14 Spaceship Earth

This project is a network of 12 neutron monitors on 4 continents designed to measure the flux of cosmic rays arriving at Earth from different directions (Figure 10.10). Ground-based neutron monitors detect variations in the approximately 500 MeV to 20 GeV portion of the primary cosmic ray spectrum. This class of cosmic ray detector is more sensitive in this portion of the cosmic ray spectrum than are cosmic ray muon detectors.

Because the intensity of cosmic rays hitting Earth is not uniform, it is important to place neutron monitors at multiple locations in order to form a complete picture of cosmic rays in space. Most of the time, a neutron monitor records galactic cosmic rays and their variation with the 11 year sunspot cycle and 22 year magnetic cycle. Occasionally the Sun emits cosmic rays of sufficient energy and intensity to raise radiation levels on Earth's surface to the degree that they are readily detected by neutron monitors. They are termed "ground level enhancements" (GLE).

Figure 10.10 Three detector tubes of the South Pole neutron monitor are on a platform in insulated, heated boxes. Data collection electronics is in "Skylab," the tall building behind. This South Pole neutron monitor is the world's most sensitive detector of relativistic solar cosmic rays, owing to its location at high latitude and high altitude (2820 meters or 9252 feet). Photo by L Shulman, 1997.

All the 12 member neutron monitor stations are located at high (Northern or Southern) latitude, which makes their detecting directions more precise, and their energy responses uniform. Their combined signals provide a real-time measurement of the three-dimensional distribution of cosmic rays, mainly galactic cosmic rays as well as solar energetic particles during the most intense solar events. Analyses of this data has applications in space weather studies.

Spaceship Earth is a multinational collaboration, with participating institutions from the United States of America, Russia, Canada, and Australia. The Bartol Research Institute at the University of Delaware, U.S.A., operates the stations at McMurdo (Antarctica), Thule (Greenland), Inuvik (Northwest Territories, Canada), Fort Smith (Northwest Territories, Canada), Peawanuck (Ontario, Canada), and Nain (Labrador, Canada).

IZMIRAN and the Polar Geophysical Institute, Russia, operate the Russian stations of Apatity, Barentsburg, Cape Schmidt, Norilsk, and Tixie Bay. The Australian Antarctic Division, Australia, operates the station at Mawson (Antarctica).

Chapter 11

11 Educational Cosmic Ray Projects

11.1 Introduction

With the great interest in cosmic rays, a number of outreach programs have been started in schools for students and teachers. Some programs discuss cosmic rays and particle physics while others have large, operating air shower arrays. These programs are located mainly on school grounds or nearby areas and are supported by larger institutions or organizations that provide the funding and technical support. These projects have been operating at various places around the world.

11.2 Cosmic-Ray Extremely Distributed Observatory (CREDO)

CREDO is a scientific project initiated in 2016 by Polish scientists from the Institute of Nuclear Physics in Krakow. Other researchers from the Czech Republic, Slovakia and Hungary also joined the project, whose purpose is the detection of cosmic rays and the search for dark matter.

Its aim is to involve as many people as possible in the construction of a global system of cosmic ray detectors, from which it will be possible to examine the essence of dark matter. Having a camera and a GPS module, as found in a smartphone, works well as a detector of particles from space.

CREDO's main objective is the detection and analysis of extended cosmic ray phenomena, so-called super-preshowers (SPS), using existing as well as new infrastructure (cosmic-ray observatories, educational detectors, single detectors etc.). They believe that the search for ensembles of cosmic ray events initiated by SPS is a new research topic and would provide a better understanding of particle physics, high energy astrophysics and cosmology.

They believe that global analysis goals are achievable when all types of existing detectors are merged into a worldwide network. The idea to use existing instruments in operation is based on a novel trigger algorithm - when looking in parallel for neighbor surface detectors receiving the signal simultaneously, one should also look for spatially isolated stations clustered in a small time window.

CREDO's strategy is also aimed at an active engagement of a large number of participants, who will contribute to the project by using common electronic devices (e.g. smartphones, Figure 11.1), capable of detecting cosmic rays. A worldwide network of cosmic-ray detectors could not only become a unique tool to study fundamental physics, but space weather or geophysics as well.

Figure 11.1 CREDO project uses smartphones to detect cosmic rays.

A user must install an application that turns their phone into a cosmic ray detector, connect it to the charger and arrange it horizontally; for example, put it on a table. Cameras must be covered with a piece of black adhesive tape, and notifications turned off. If a radiation particle passes through a photosensitive matrix in the phone, it will stimulate several pixels, which will be noticed by the program that sends information to the server. Time and place of the event is also known, thanks to the GPS module. Data from the smartphones will then be analyzed together in the Academic Computer Center Cyfronet AGH, which will keep participants informed about the progress of the search for signs of high-energy particles. As of 2021 this project was still active.

11.3 HiSparc

HiSparc (High School Project on Astrophysics Research with Cosmics) is a very successful outreach project in which secondary school students participate in real research on cosmic rays. It has been running for more than 10 years, with participants in Holland Germany, Denmark, Austria, Vietnam, Poland and the UK. Participating schools receive a DIY detector kit, which students build and install at their school for collecting data.

HiSparc students contribute to actual scientific research and to the understanding of cosmic rays. Schools receive a self-assembly cosmic ray detector kit that helps students learn about cosmic rays as well as collect data to analyze. Detector data is fed into a shared data base for use by researchers. Institutions across the globe have access to this data.

Students gain experience by assembling and installing the DIY cosmic ray detector kit at their school (Figure 11.2). They can choose between a two-detector or four-detector system. Kits contain everything needed for installation, including cables and materials to fix it to the roof.

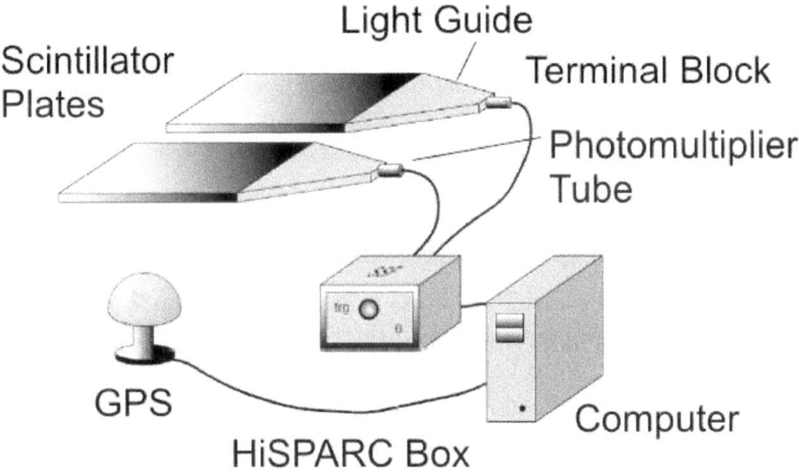

Figure 11.2 HiSparc cosmic ray detector using scintillator panels.

HiSparc central can remotely log in to check the detector's performance and to diagnose problems and will notify schools if there is an issue. They also ensure the DAQ software and firmware is automatically updated remotely. The only thing schools need to provide is power, a computer and an internet connection. HiSparc maintains a website showing all the participants and current data.

11.4 Cosmic@Web

This is a very unusual approach to studying cosmic rays. Students do not build or operate cosmic ray detectors. Yet, without any programming knowledge and working comfortably from home with a laptop computer, pupils can perform a variety of experiments, most of which are carried out at DESY in Zeuthen, Germany. They are integrated into Cosmic@Web and thus offer easy access to real long-term measurements.

Having an experiment on-site is not always possible in scientific research, particularly since large-scale experiments for particle and astroparticle physics are often complex and expensive. DESY (Deutsches Elektronen-Synchrotron) participates in such projects as the IceCube experiment in Antarctica, experiments at the Large Hadron Collider (LHC) at CERN and the planned Cherenkov Telescope Array (CTA).

To expand the possibility of investigations with cosmic particles in the classroom and to reach a broader audience, the use of experimental data by students via the internet was introduced. DESY provides the internet portal Cosmic@Web which allows students to analyze a large amount of data taken by different cosmic particle experiments running continuously at DESY, on the research vessel Polarstern and at the Antarctic station Neumayer III. A detailed description of the experiments, of the data structure, a proposal of interesting problems for students to investigate and a tutorial on how to use the web interface allows independent work. Students can analyze the data from the classroom or from home without direct contact to scientists.

11.5 Extreme Energy Events (EEE)

EEE project (Science inside schools) is a joint educational and scientific initiative for studying cosmic rays that was conceived in 2003. This strategic project of Centro Fermi, Rome is conducted in collaboration with CERN, INFN and MIUR and carried out with the essential contribution of high school students and teachers.

Physics research interests include the properties of the local muon flux, the detection of extensive air showers, and the search for possible long range correlations between far telescopes. Experiments are based on a network of "telescopes" using the most advanced particle detectors (Multi-gap Resistive Plate Chambers, MRPC) built at CERN by teams of students and teachers (Figure 11.3).

Each station of the EEE network, is made of three Multigap Resistive Plate Chambers (MRPC) specifically designed to achieve good tracking and timing capability, low construction costs, and an easy assembly procedure. A single MRPC is shown in the figure. Three such MRPC chambers are placed one above the other with the top and the bottom chambers at a distance of 50 cm from the middle chamber. This arrangement provides tracking of muons as they proceed through the chamber.

Telescopes are located in high schools distributed throughout Italy and are controlled by students. Currently, about 50 telescopes are taking data, and more than 90 institutes are analyzing data. Data from all telescopes are centrally collected, reconstructed and distributed to the students. Regular videoconferences, master classes, meetings and visits are organized with the involvement of all institutes. More than 50 billion tracks have been collected and are presently studied by students and professional researchers. The project is expanding with the construction of new telescopes. Website: EEE

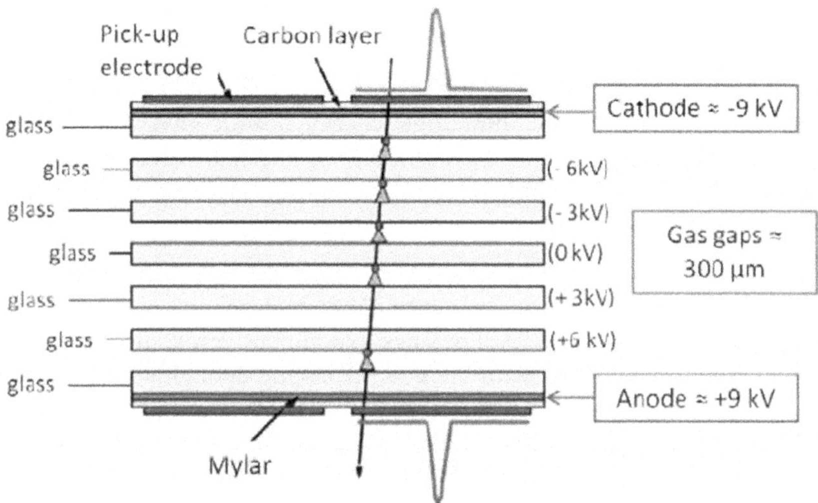

Figure 11.3 An EEE single Multi-gap Resistive Plate Chamber (MRPC).

11.6 QuarkNet

QuarkNet began in 1999 as a long-term, research-based teacher professional development program in the United States jointly funded by the National Science Foundation and the US Department of Energy. It has established centers at universities and national laboratories conducting research in particle physics across the United States.

The main QuarkNet student investigations supported at the national level are cosmic ray studies. Working with Fermilab technicians and research physicists, QuarkNet staff has developed a classroom cosmic ray muon detector that uses the same technologies as the largest detectors at Fermilab and CERN.

To support interschool collaboration, QuarkNet collaborates with the Interactions in Understanding the Universe Project (I2U2) to develop and support the Cosmic Ray e-Lab. An e-Lab is a student-led, teacher-guided investigation using experimental data. Students have an opportunity to organize and conduct authentic research and experience the environment of a scientific collaboration.

Participating schools set up a detector somewhere at the school. Students collect and upload the data to a central server located at Argonne National Laboratory. Students can access data from all of the detectors in the cluster and use this data for studies, such as determining the (mean) lifetime of muons, the overall flux of muons in cosmic rays, or a study of extended air showers.

11.7 Washington Area Large-scale Time-coincidence Array

WALTA is a cosmic ray physics experiment run by the University of Washington to investigate ultra high energy cosmic rays ($>10^{19}$eV). It began around 2003 and uses detectors placed at Seattle-area high schools and colleges which are linked via the internet, effectively forming an Extensive Air Shower array.

In addition to detecting cosmic rays it hopes to serve as a pedagogical tool for increasing the physics involvement of high schools and community colleges with a University level physics experiment. Each site has three to four scintillation detectors with the goal of having enough sites to cover a 200 km^2 area around the city of Seattle. WALTA is a part of the larger NALTA project which hopes to combine data from several WALTA like projects to further the exploration of UHE cosmic rays.

WALTA's goal was to set up detectors at 32 sites in the Seattle area, covering an area of 200 square kilometers. This area would be large enough to detect events above the GZK cutoff. There has not been much activity lately on the WALTA web site and it has not been updated recently.

11.8 California High School Cosmic Ray Observatory (CHICOS)

This cosmic ray observatory was operated by the Kellogg Laboratory at the California Institute of Technology in Pasadena, California. It began around 2001, with a grant by the National Science Foundation to CalTech. It was designed to be one of the world's largest ongoing Cosmic Ray observatory programs. It was known for its large network within the Los Angeles County area, based mainly on high school sites carrying cosmic ray detector units, though there were also detector units on associated elementary schools and middle schools.

The CHICOS project represented a unique blending of cutting-edge scientific research and broadly-based educational outreach. It employed an active research array for the detection of ultra-high-energy cosmic rays. The deployed array has already seen extended air showers. There has not been much activity lately on the CHICOS web site and it has not been updated recently.

11.9 Showers of Knowledge

This is an open outreach educational project in Russia that aims to permit internet users worldwide to analyze data of real online cosmic-ray experiments. It began in 2013 and was developed at the Joint Institute for Nuclear Research (Dubna, Russia).

The project consists of a distributed setup for researching cosmic rays RUSALKA ("mermaid"), comprising 11 stations located in an area of about 0.5 km in diameter. It uses the interactive internet portal *livni.jinr.ru*, where users can run a variety of pre-made data analysis scripts with their custom parameters. It features the possibility for users to communicate with real particle physicists developing the project.

Each individual working station consists of two scintillation counters, a receiver for a global positioning system GPS or GLONAS and a block of electronics for forming and inputting data to the computer that has access to the Internet.

A typical RUSALKA scintillation counter consists of two plates of plastic scintillator. Thickness of each plate is 2 cm and dimensions 30 cm x 80 cm, so the total sensitive area of the counter is 60 cm x 80 cm. Light is collected by a set of wavelength shifting optical fibers. Optical fibers guarantee transfer of light to photomultiplier tubes (PMT) with minimum losses. In the RUSALKA counters PMTs FEU-85 (USSR) or Hamamatsu (Japan) are used.

Access to data, continuously obtained by such a setup, permits a participant from the very beginning to take part in studying the properties of cosmic rays and becoming familiar with modern scientific and practical problems - starting with hypotheses concerning development of the Universe and ending up with methods of handling and transferring data via the Internet.

Chapter 12

12 Importance of Cosmic Rays to Our Planet

12.1 Introduction

Cosmic rays, like many things in science, seem distant from our everyday lives. Yet, cosmic rays *do* create effects that touch our lives and our planet. These effects are due to the primary solar and galactic rays that impact Earth. These protons and alpha particles penetrate our atmosphere and the surface with significant outcomes.

The best-known example is dating by radiocarbon, or carbon-14 which is produced by cosmic rays in the Earth's atmosphere and enters every living animal and plant. Archaeological dating has benefitted greatly from measurement of ^{14}C in organic remains like wood and skeletons.

At sea level we are continually irradiated by secondary CR particles and receive an unavoidable radiation dose. These effects tend to grow in strength with an increase in altitude. Thus passengers on high altitude planes or astronauts might receive severe radiation doses during large solar flares.

Recently it has been suggested that the Earth's cloud cover is correlated with the intensity of galactic cosmic rays. The idea is that droplet formation is influenced by ionization of atmospheric molecules by cosmic rays. We'll delve more into these subjects in the following sections.

12.2 Carbon 14

Earth's chronological history has fascinated civilized humanity for thousands of years. How old is the Earth? Where and when were humans created? Myths and legends gave the only answers until scientific methods and the discovery of radioactivity changed it.

Earth's age is now firmly established based on measurements of lead isotopes as well as radioactive uranium and thorium whose proportions change measurably over billions of years. But human history and archeology look at much shorter time spans in which our societies developed and left numerous relics. At first, dating of buildings, skeletons and other artifacts was established from clues such as historical records, legends and the strata of the investigation site.

In 1946 Willard Libby, of the University of Chicago, proposed an innovative method for dating organic materials by measuring their content of carbon-14, a newly discovered radioactive isotope of carbon. With this dating technique he initiated a revolution in the fields of archaeology and geology by allowing practitioners to develop more precise historical chronologies across geography and cultures.

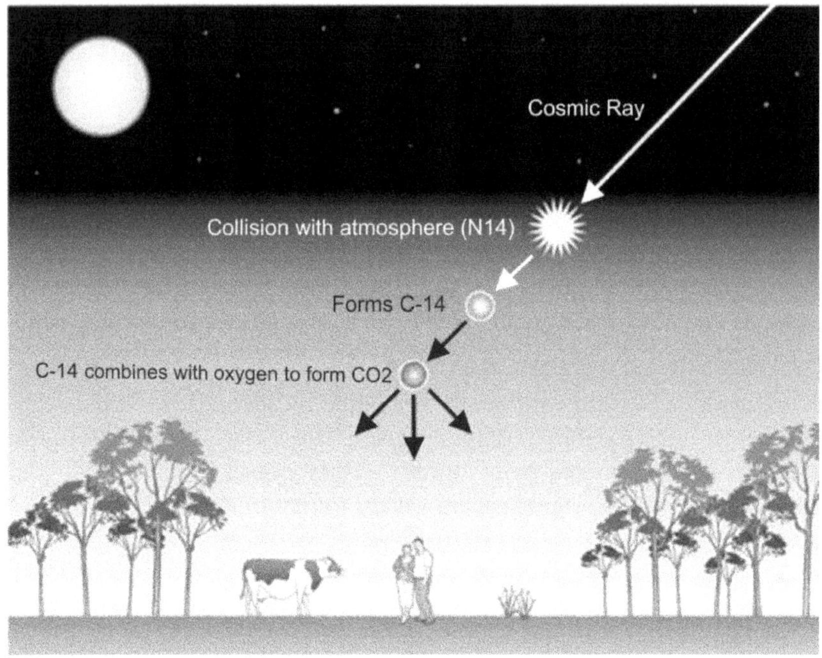

Figure 12.1 Generation of CO2 from cosmic rays, which enters our biosphere.

Libby was inspired by physicist Serge Korff of New York University, who in 1939 discovered that neutrons were produced during the bombardment of the atmosphere by cosmic rays. Korff predicted that the reaction between these neutrons and nitrogen-14, which predominates in the atmosphere, would produce carbon-14 (Figure 12.1).

This newly created ^{14}C is radioactive, with a half-life of 5,730 years. It decays back to nitrogen after emitting an electron. The latent radioactivity does not affect the chemical ability of the carbon-14 to combine with other atoms to form molecules. The radiocarbon combines with oxygen and produces radioactive carbon dioxide, $^{14}CO2$, which then spreads through the biosphere.

These molecules attach to tiny dust particles which follow air currents, becoming incorporated in rain. Some carbon dioxide is absorbed by plants that are eaten by humans and other animals. And some goes into rivers and then into oceans. Spreading of CO2 in the atmosphere, plants, animals, and oceans is relatively fast compared to the half-life of carbon-14

Libby's concept of radiocarbon dating focused on measuring the carbon content of discreet organic objects. But in order to prove the idea he needed to understand the Earth's carbon system. Radiocarbon dating would be most successful if two important factors were true: that the concentration of carbon-14 in the atmosphere had been constant for thousands of years, and that carbon-14 moved readily through the atmosphere, biosphere, oceans and other reservoirs—in a process known as the carbon cycle.

In the absence of any historical data concerning the intensity of cosmic radiation, Libby simply assumed that it had been constant. He reasoned that a state of equilibrium must exist wherein the rate of carbon-14 production was equal to its rate of decay, dating back millennia. Fortunately his assumptions turned out to be generally true.

Regular carbon that we find on Earth consists almost entirely of a mixture of two isotopes: 98.9 percent is 12C (six protons and six neutrons in each nucleus), and the other 1.1 percent is 13C, which has an additional neutron in each nucleus. Living things have an average concentration of one carbon-14 atom for every trillion regular carbon atoms. This is the concentration existing in all living things. When a plant or an animal dies, its intake of carbon dioxide with its carbon-14 ceases. Thereafter, the ratio 14C / (12C + 13C) changes slowly as carbon-14 decays. Determining this ratio enables us to estimate the age of the wood, skeletons, paper, or cloth.

An equilibrium concentration of radiocarbon shows up in the emission of 13.5 electrons per minute for each gram of carbon in a fresh sample. In wood that is 2,000 years old about 79 percent of the radiocarbon will have survived and the decay rate will be about 10 counts per minute. By 4,000 years, the corresponding rate will be about 8 counts per minute.

Measuring such small concentrations of carbon-14 is not easy. Libby's first detector was a Geiger counter of his own design. He converted the carbon in his sample to lamp black (soot) and coated the inner surface of a cylinder with it. This was inserted into the counter to detect electrons emitted by decaying ^{14}C. A far more sensitive technique for carbon-14 detection was developed in the 1970s. It identifies atoms by their masses rather than by their radioactivity. A minute sample is vaporized and the fully ionized atoms are sent through a cyclotron or other type of accelerator where carbon-14 is separated from carbon-12 and carbon-13.

12.3 Biological Effects

When cosmic rays and other high-speed particles travel through living tissue, some of the particles' kinetic energy is transferred to the tissue, where electrons can be ejected from atoms. When the electron that was shared by the two atoms to form a molecular bond is dislodged by the ionizing radiation, the bond is broken and thus, the molecule falls apart. This is a basic model for understanding radiation damage. When ionizing radiation interacts with cells, it may or may not strike a critical part of the cell.

Chromosomes are considered to be the most critical part of the cell since they contain the genetic information and instructions required for the cell to perform its function and to make copies of it for reproduction purposes. Obviously, cosmic radiation has been an intrinsic condition for the evolution of life on Earth since its very beginnings. Nature, therefore, had to develop efficient repair mechanisms, which are able to repair cellular damage –including chromosome damage.

A measure of the health effect of low levels of ionizing radiation on the human body is the derived unit of ionizing radiation called the sievert (Sv). It expresses the dose equivalent (the biological effect) of the radiation, associated with all possible exposure situations on the human body. One sievert carries with it a 5.5% chance of eventually developing fatal cancer. A sievert is a large value so usually milli-sieverts (mSv) are commonly used. Typical values for yearly background radiation at sea level are: inhalation of air 2.28 mSv, ingestion of food and water 0.25 mSv, terrestrial radiation 0.21 mSv and cosmic radiation 0.33 mSv.

The radiation we observe at aircraft altitudes of 12 km (39,000 feet) is due to very high-energy particles - mainly protons and helium nuclei, together with a small amount of heavy nuclei - penetrating the atmosphere and colliding with air atoms. For airline crew members, on the average at flight altitude, radiation dosage is of the order of 10 microSv / hour. This value varies with altitude, latitude, and solar activity and it must be interpreted in comparison with the average natural dosage. Thus, approx.400 hours per year at flight altitude would lead to the equivalent of the natural yearly radiation load of around 4.0 mSv.

Astronauts are subjected to the full intensity of high-energy cosmic rays and solar particles (together with the secondary particles produced in the spacecraft walls), and the biological risks in space are the subject of ongoing investigations. A typical mission to Mars, for example, could result in a total "dose equivalent" of up to 500 mSv. Current estimates suggest that a person who receives a 1-4 Sv dose of ionizing radiation incurs a few per cent increase in the risk of contracting fatal cancer in their lifetime, although the risk level depends on sex and age.

12.3 Weather and Climate

Cosmic rays effects associated with weather and climate is currently a subject of intense study and controversy. Here are some possible roles of cosmic rays in these areas.

Potential Role in Lightening. Cosmic rays have been implicated in the triggering of electrical breakdown in lightning. A mechanism proposed by Russian physicist Alex V. Gurevich of the Lebedev Physical Institute suggests that the movement of large showers of energetic particles produced by high-energy cosmic rays might provide a conductive path that initiates lightning. There are types of particle detectors called spark chambers that exploit this principle. In a spark chamber, a very large voltage is applied across a small gap filled with a gas. If the electric field is large enough and there are some free electrons available, the gap breaks down (spark) when a charged particle passes through.

For thunderstorms and lightning the situation is slightly different. Unlike the spark chamber, the electric fields inside the thunderstorm do not appear to be big enough to initiate a spark. In order for Gurevich's mechanism to work, he supposed that there were many, many charged particles passing through the storm at once. Because cosmic-ray air showers do not produce enough particles by themselves, Gurevich postulated that the thunderstorm gave the cosmic-ray shower a boost by increasing the number of energetic electrons through an exotic process called "runaway breakdown."

Runaway breakdown occurs when the drag force electrons experience moving through air is less than the electric force acting upon them. In such cases, the electrons will "run away," gaining very large amounts of energy. As the runaway electrons collide with air molecules, they generate other runaway electrons plus x-rays and gamma rays, resulting in an avalanche of high-energy particles.

Runaway breakdown can create large amounts of high-energy electrons, as well as x-rays and gamma rays. Interestingly, we know that runaway breakdown works for the low electric fields already seen inside thunderstorms. We also know that it does sometimes happen right before lightning, because we can see big bursts of x-rays and gamma rays shooting out of thunderstorms. But we do not know for sure that the mechanism described above is at work in our atmosphere.

Postulated Role in Climate Change. We normally associate changes in the Earth's climate to internal causes like volcanic dust in the atmosphere, atmospheric/ocean oscillations like the El Niño Southern Oscillation, and to external causes like variations in the Earth's orbital parameters and in the Sun's luminosity. But some scientists believe that cosmic rays may have influenced our climate.

Back in 2007 Adrian Melott and Mikhail Medvedev (University of Kansas) reported that 62-million year cycles in biological marine populations correlate with the motion of the Earth relative to the galactic plane and increases in exposure to cosmic rays. The researchers suggest that this and gamma ray bombardments deriving from local supernovae could have affected cancer and mutation rates, and might be linked to decisive alterations in the Earth's climate, and to the mass extinctions of the Ordovician.

In recent years, Danish physicist Henrik Svensmark has controversially argued that because solar variation modulates the cosmic ray flux on Earth, they would consequently affect the rate of cloud formation and hence be an indirect cause of global warming. Other scientists have vigorously criticized Svensmark for inconsistent work: one example is adjustment of cloud data that understates error in lower cloud data, but not in high cloud data; another example is "incorrect handling of the physical data" resulting in graphs that do not show the correlations they claim to show. Despite Svensmark's assertions, galactic cosmic rays have shown no statistically significant influence on changes in cloud cover, and have been demonstrated in studies to have no causal relationship to changes in global temperature.

In 2019, Professor Masayuki Hyodo (Research Center for Inland Seas, Kobe University) reported new evidence suggesting that galactic cosmic rays affect the Earth's climate by increasing cloud cover, causing an "umbrella effect". He noted that when galactic cosmic rays increased during the Earth's last geomagnetic reversal transition 780,000 years ago, the umbrella effect of low-cloud cover led to high atmospheric pressure in Siberia, causing the East Asian winter monsoon to become stronger. This is evidence that galactic cosmic rays influence changes in the Earth's climate.

As these hypotheses are still controversial and speculative they are currently the objects of intense world-wide research activities.

12.4 How Cosmic Rays May Have Shaped Life.

In 2020 Noémie Globus and others at Stanford University proposed that that the influence of cosmic rays on early life may explain nature's preference for a uniform 'handedness' among biology's critical molecules. They speculate that this interaction between ancient proto-organisms and cosmic rays may be responsible for a crucial structural preference, called chirality, in biological molecules. If their idea is correct, it suggests that all life throughout the universe could share the same chiral preference.

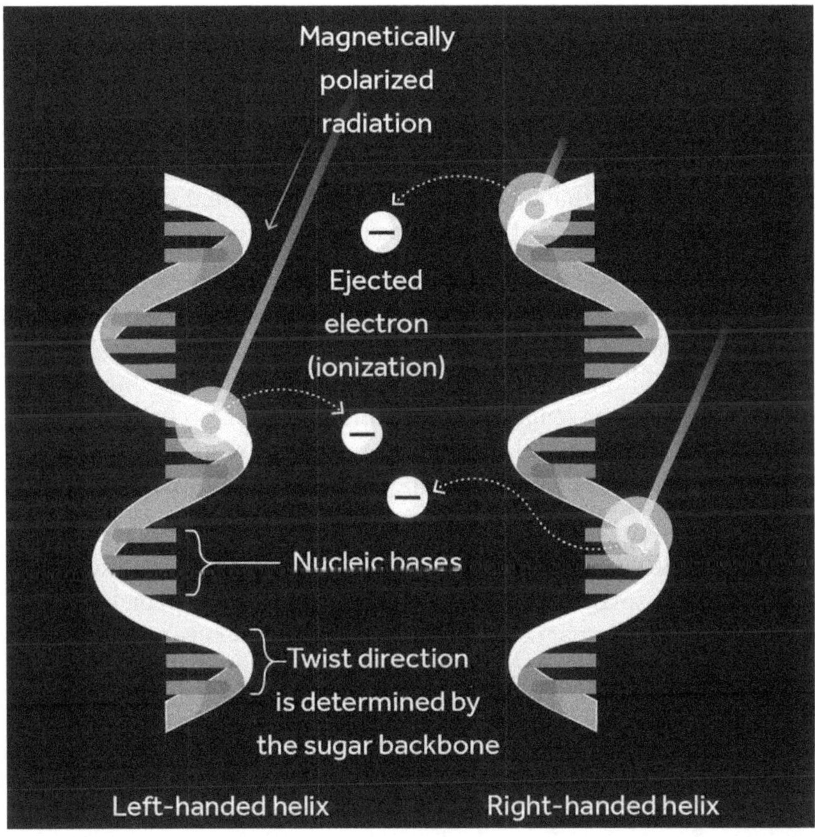

Figure 12.2 Magnetically polarized radiation preferentially ionized one type of "handedness" leading to a slightly different mutation rate between the two mirror proto-life forms. Over time, right-handed molecules out-evolved their left-handed counterparts. (Image credit: Simons Foundation)

Chirality, also known as handedness, is the existence of mirror-image versions of molecules (Figure 12.2). Like the left and right hand, two chiral forms of a single molecule reflect each other in shape but don't line up if stacked. In every major biomolecule -- amino acids, DNA, RNA -- life only uses one form of molecular handedness. If the mirror version of a molecule is substituted for the regular version within a biological system, the system will often malfunction or stop functioning entirely. In the case of DNA, a single wrong handed sugar would disrupt the stable helical structure of the molecule.

Louis Pasteur first discovered this biological homochirality in 1848. Since then, scientists have debated whether the handedness of life was driven by random chance or some unknown deterministic influence. Pasteur hypothesized that, if life is asymmetric, then it may be due to an asymmetry in the fundamental interactions of physics that exist throughout the cosmos.

Globus proposed "that the biological handedness we witness now on Earth is due to evolution amidst magnetically polarized radiation, where a tiny difference in the mutation rate may have promoted the evolution of DNA-based life, rather than its mirror image,"

The researchers base their argument on cosmic ray muons that are formed near the top of the atmosphere from primary rays. They believe the muons are magnetically polarized, meaning, on average, muons all share the same magnetic orientation. When muons finally decay, they produce electrons with the same magnetic polarization. The researchers believe that the muon's penetrative ability allows it and its daughter electrons to potentially affect chiral molecules on Earth and everywhere else in the universe.

Globus explained that "We are irradiated all the time by cosmic rays," "Their effects are small but constant in every place on the planet where life could evolve, and the magnetic polarization of the muons and electrons is always the same. And even on other planets, cosmic rays would have the same effects."

The researchers' hypothesis is that, at the beginning of life on Earth, this constant and consistent radiation affected the evolution of the two mirror life-forms in different ways, helping one ultimately prevail over the other. These tiny differences in mutation rate would have been most significant when life was beginning and the molecules involved were very simple and more fragile. Under these circumstances, the small but persistent chiral influence from cosmic rays could have, over billions of generations of evolution, produced the single biological handedness we see today.

Globus and co-author Blandford suggest experiments that could help prove or disprove their cosmic ray hypothesis. For example, they would like to test how bacteria respond to radiation with different magnetic polarization. "Experiments like this have never been performed and I am excited to see what they teach us. Surprises inevitably come from further work on interdisciplinary topics," said Globus.

Chapter 13

13 Applications of Cosmic Rays

13.1 Introduction

When cosmic rays strike Earth's atmosphere, a cascade of many types of subatomic particles is created. Hadronic particles produced in the shower either decay or interact with other particles, while electrons and photons lose energy rapidly through pair production and Bremsstrahlung so that by the time this shower of particles reaches the Earth's surface, it is comprised primarily of muons. The flux at the surface of the Earth is about 10,000 muons per minute per square meter. At sea level, the mean energy of muons is about 3–4 GeV, sufficient to penetrate meters of rock. Making practical use of these energetic free particles is an attractive idea.

In the 1950's E.P. George pioneered the use of cosmic ray muons for imaging purposes. He employed a Geiger counter muon telescope to determine the depth of the ice burden above the Guthega–Munyang tunnel in Australia.

In the late 1960,s a team led by Nobel-prize-winning physicist Luis Alvarez used muon transmission imaging to look for hidden chambers in the Pyramid of Chephren in Giza (Figure 13.1). He was an expert in particle detectors and used spark chambers as detectors and recorded the cosmic-ray data on magnetic tape. He demonstrated that no such chambers were present, which was of considerable significance. This showed that this technique can be used to "screen" pyramids, in order that only those with apparently interesting internal features are chosen for more detailed physical exploration by archaeologists, the others being left alone.

These first applications were based on the measurement of the attenuation of the cosmic ray flux. By counting the number of muons exiting the volume under inspection and assuming the composition of the material crossed, the total thickness of the material could be inferred. Clearly the number of surviving muons is related to the type and thickness of the interposed material. This technique has been reutilized and improved more recently to visualize magma dynamics in an erupting volcano.

Other interesting and unusual applications have also been proposed and realized such as imaging of large vessels, cosmic muon detection for geophysical applications and observation of the moon's shadow.

Figure 13.1 The birth of the idea. Luis Alvarez invented muon imaging in the 1960's to study the 2nd Pyramid of Chephren.

In 2002 a break-through idea was proposed by the Department of Energy's Los Alamos National Laboratory. Instead of using the absorption or transmission properties of cosmic ray muons, the scattering angle of the surviving muons could be used to infer the properties of the material crossed. This novel technique opened up new opportunities and indeed more groups around the world began investigating the possibility of applying this method for civil applications.

The properties of cosmic ray muons define many useful applications. They are highly penetrating, their average energy is about 10 000 times the energy of a typical X-ray and they can penetrate hundreds of meters of rock. This means that they are suitable to image objects that are behind or inside of shielding material that is too thick or deep for other imaging methods. They are natural and everywhere, which means that any imaging method using them is entirely passive and therefore neutral with regards to health and safety regulations. But their low flux means that they are suitable only for applications that are not time-critical. Finally, cosmic-ray muons are cost-free and their availability is unlimited. This makes them an ideal radiation source.

In this chapter we use the terms 'muon radiography' and 'muon tomography' for two dimensional absorption imaging and for three-dimensional imaging similar to x-ray tomography, respectively. The term 'muography' has been established in the community of volcanologists that use cosmic-ray muons to image volcanoes as is the term 'muon imaging'. In the remainder of this chapter we will discuss the technology of muon detection and look at some of these new muon applications.

13.2 Muon Detector Systems

There are several types of instruments that can detect muons. The ones most commonly used for tracking muons include scintillation counters and gas wire detectors such as drift tubes and cathode strip chambers.

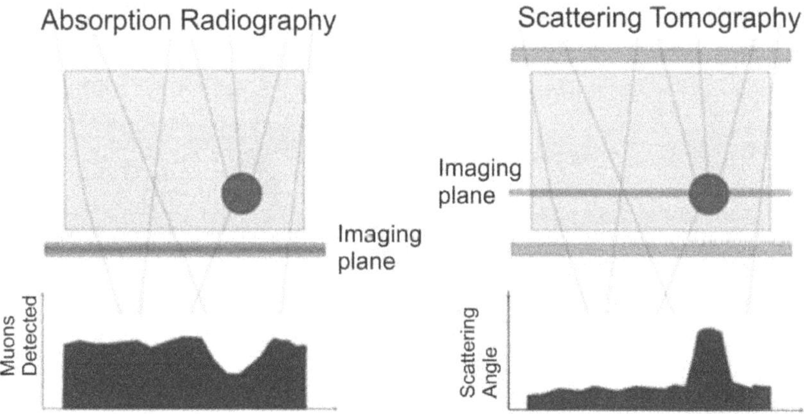

Figure 13.2 Schematic visualization of absorption radiography and scattering tomography for a solid sphere and the corresponding detector geometries.

There are three common types of detector systems: mobile radiography systems, static tomography systems and borehole detectors. Figure 13.2 illustrates typical detector systems and imaging techniques used for absorption radiography and scattering tomography.

Muon radiography requires at least two detector planes in order to define the tracks of the detected cosmic muons to produce a two-dimensional absorption image. Often three or four detector planes are used for better resolution and efficiency. Muon radiography results are not necessarily limited to two-dimensional images. Information from several detectors imaging the same volume can be combined to form a three-dimensional image

Muon tomography requires at least two detector planes above and two below the object that is to be imaged. The upper detectors can be seen as defining the radiation source, similar to the x-ray source in a CT system, while the lower detectors detect the presence, absence and scattering of the muons that were defined by the upper detectors. This is, in fact, very similar to a CT system, but due to the angular distribution of cosmic muons and due to the geometric limits imposed by the detector system only a narrow angular range around the vertical is covered. This means that the resolution of such a muon tomography system in the horizontal plane (in x and y) will be much better than the vertical resolution along the z-axis.

Figure 13.3 The Los Alamos Mini Muon Tracker. Two identical muon tracking modules are used to determine the scattering angle of muons that pass through objects placed in between them.

Existing muon tomography systems are static systems requiring that the object or sample be placed inside their active volume (Figure 13.3). This active volume is typically of the order of one cubic meter. Larger ones may have ten times that volume.

13.3 Imaging the Core of Fukushima Reactor with Muons

A nuclear crisis at the Fukushima Daiichi reactors resulted from a 9.0 magnitude earthquake, followed by the vast tsunami on March 11, 2011.

Damage of the reactor cores attracted worldwide attention to the issue of the safety of nuclear energy. After a cold shutdown in December 2011 the Japanese government began a new phase of cleanup and decommissioning. However, it is difficult to plan the dismantling of the reactors without a realistic estimate of the extent of the damage to the cores and knowledge of the location of the melted fuel. High radiation fields permit only limited access to the reactor buildings (Figure 13.4).

Muon imaging offers the potential to image the nuclear reactor cores without access to the cores. It utilizes naturally occurring cosmic-ray muons to image dense objects. Los Alamos National Laboratory researchers developed a sensitive technique, called muon scattering radiography. This scattering method uses two muon trackers to measure incoming and outgoing tracks of individual muons, where the region of interest is contained within the volume of the tracker pair. Combining the incoming and outgoing muon tracks provides good spatial resolution.

Figure 13.4 Damaged Unit-1 reactor of Fukushima Daiichi, viewed from the north-west side. The Unit-2 building is seen to the right. (Image TEPCO)

The scattering method has high sensitivity to high-Z materials such as the uranium in nuclear fuel, making it useful for detecting uranium in a back-ground of low-Z material. The researchers determined that it may be feasible to image the Fukushima Daiichi reactors with cosmic-ray muons to assess the damage to the reactors. Their method has high sensitivity for detecting uranium fuel and debris even through thick concrete walls and a reactor pressure vessel.

In 2011 two teams of physicists started building muon detectors specifically for Fukushima Daiichi. Japanese researchers at the High Energy Accelerator Research Organization (known as KEK) constructed a system that uses muon transmission imaging, the same type of system used in the Egyptian pyramid. In a separate effort, Toshiba, a major contractor for Tokyo Electric Power Co (TEPCO), asked Decision Sciences International Corporation (DSIC) to build an instrument based on its muon-scattering system licensed from Los Alamos.

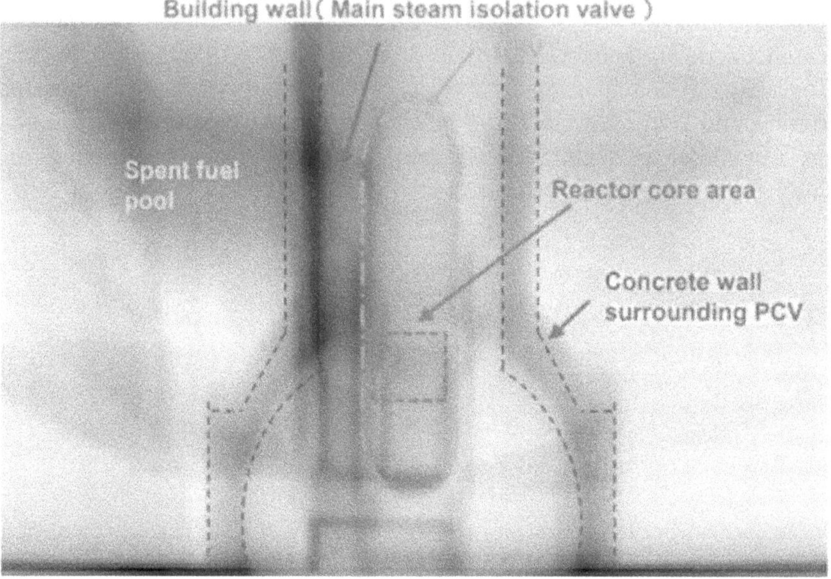

Figure 13.5 Structures within the reactor building of unit 3 can be seen in images obtained using muon data. (Image TEPCO)

In February 2015, TEPCO debuted the KEK system at reactor 1, setting up the detector outside the reactor building and letting it collect muon data for 90 days. TEPCO says the images verified its working assumptions about the behavior of reactor 1's fuel load - namely, that all of it melted and flowed downward. The problem is that the images reveal the interior of the reactor building only in broad strokes (Figure 13.5). Engineers say it's hard to imagine that these fuzzy images will help make serious decisions about defueling operations.

Meanwhile, Toshiba and the International Research Institute for Nuclear Decommissioning (IRID) have adapted the technology for measuring the scattering behavior of muons penetrating objects, building on techniques originally developed by the Los Alamos National Laboratory (LANL). Two external detectors are placed in parallel to one another on opposite sides of the reactor building, and cosmic-ray muon trajectories, scatter and scatter angle are recorded (Figure 13.6). The scatter angle is directly related to the atomic number of objects the muons collide with, and uranium can be distinguished from surrounding structural objects with lower atomic numbers.

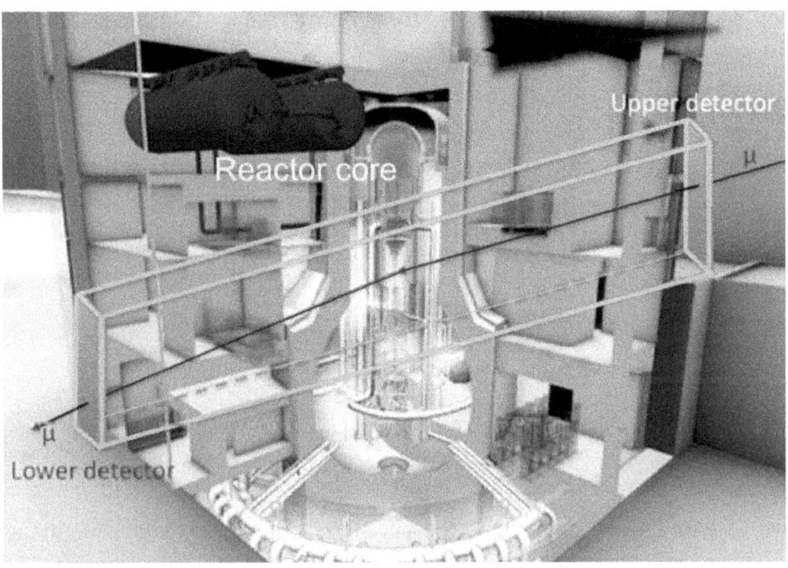

Figure 13.6 Arrangement of two scattering muon detectors and detecting volume in a model of Fukushima nuclear power plant.

In 2014 Decision Sciences was awarded a contract by Toshiba Corporation Power Systems to deliver a non-standard configuration of their Multi-Mode Passive Detection System MMPDS to safely image the cores of the nuclear reactors at the Fukushima Daiichi Nuclear Power Plant that were disabled during the March 2011 nuclear disaster.

Currently the Decision Sciences system sits in storage at a Toshiba research center in Japan, where it has been calibrated and could be used whenever TEPCO wishes. The system has already been tested on a small-scale mock-up of a Fukushima reactor at Los Alamos, and on a working research reactor at the Toshiba facility. Those tests appeared good.

Yet the device wasn't tested at Fukushima then, and instead still waits in storage. At this time, TEPCO won't specify a timeline for its use. As of 2016, a TEPCO spokesman would say only that the KEK device will be given priority at reactor 2, and that the Decision Sciences system will be installed "depending on the status of on-site work."

13.4 Cargo Scanning with Muons

In 2021 a team of researchers in Italy and the US, led by Francesco Riggi at the University of Catania reported progress on a full-scale prototype muon tomograph that can peer inside cargo containers. They have combined layers of muon detectors with a specialized reconstruction algorithm to deliver high-resolution 3D images of a small lead block inside a large sensing area. This technology could make it far easier for cargo authorities to stop dangerous nuclear materials from being transported illegally.

Figure 13.7 Scientists in Italy and the US are developing detector systems using cosmic muons to peer inside the steel boxes.

Cargo containers are widely used to transport high volumes of goods and because of their robust steel construction and large size, small objects can easily be concealed inside them (Figure 13.7). As a result, there is a growing concern that containers could be used to transport illicit nuclear materials, including plutonium and uranium.

The researchers have built a full-scale prototype (18 m^2 sensitive area) of a muon tomograph with all features to be potentially employed for the inspection of a standard Twenty-Foot Equivalent (TEU) container.

Riggi and colleagues have combined several techniques to create a full-scale muon tomograph. Their setup features multiple layers of scintillator-based muon detectors positioned above and below the sensing area. These detectors track changes in the paths of muons as they are scattered by dense materials. Then, an algorithm analyses the muon trajectories to estimate the point of closest approach between the muons and heavy atomic nuclei. From this information, they can create high-resolution 3D images of dense materials inside the sensing area (Figure 13.8).

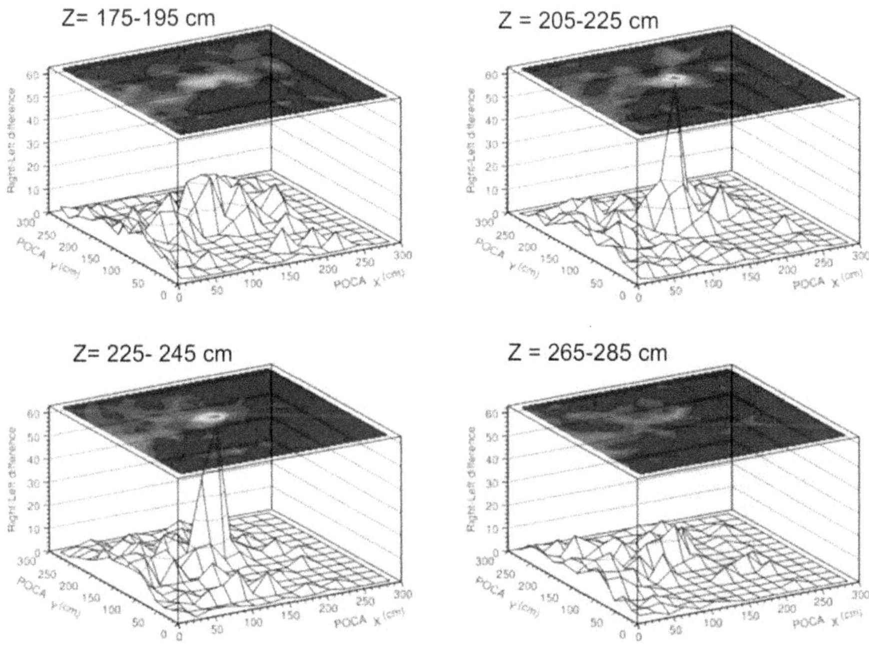

Figure 13.8 These are 2D xy maps for various z levels of a small 20 cm lead block target positioned at a height of z =225 cm in the detection chamber.

Riggi's team was able to precisely determine the 3D position of a small lead block about 20 cm across within a sensing cross-sectional area of 18 m^2, which is large enough to accommodate a standard cargo container. The team says that their successful demonstration paves the way for devices that can efficiently detect concealed nuclear materials. With further improvements to reduce scanning times, the inspection devices may one day become an integral part of cargo handling facilities worldwide.

Other multi-mode passive detection systems based upon muon tomography, are currently in use by Decision Sciences International Corporation at Freeport, Bahamas, and the Atomic Weapons Establishment in the United Kingdom.

13.5 Decision Sciences Freeport Muon Inspection Station

Decision Sciences has a MMPDS Gen2 system located at the Hutchinson Port Holdings Terminal at the Freeport Container Port in Freeport, Bahamas. This system is used system primarily for the Nuclear and Radiological

Imaging Program of the U.S. Department of Homeland Security. Installed in 2012, it has scanned several thousand vehicles, most of them shipping container trucks. The system has maintained >95% operational up time for over three years, even in the face of tropical storms (Figure 13.9).

Figure 13.9 The MMPDS Gen2 in Freeport, Bahamas. One container truck is being scanned while another truck immediately behind it approaches the MMPDS for its turn. Local control and computer rooms are in the small building to the right.

Decision Sciences' now has the MMPDS Gen3 with improved capability that also incorporates Machine Learning Algorithms (MLAs) that will steadily improve the capability of the system against current items of interest and rapidly adapt to future threats as they emerge.

When a vehicle or container to be inspected is positioned inside the MMPDS, naturally occurring cosmic ray particles (consisting of muons and electrons) pass through both the detector and the vehicle or container. During the scan the detector measures the trajectory of these particles before and after they pass through the vehicle or container. Any deviations to the trajectory of the particles including scattering and absorption or attenuation caused by the items inside a container or vehicle, are captured, measured, and analyzed by the MMPDS hardware, electronics, and highly sophisticated system software.

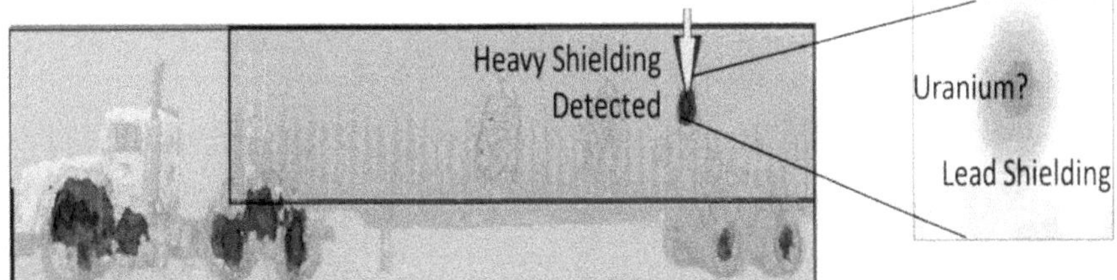

Figure 13.10. Operator view (left) with horizontal-plane and vertical-plane (right) slices through a 3-D tomographic muon tomography image, with superimposed spheres showing location of a radioactive source estimated via gamma ray source localization. The muon tomography image shows the true location of the source.

Decision Sciences' proprietary MLAs use the signature profiles and features of the 3-D image map to identify objects that may be nuclear or radiological threats or shielding enclosures which can stop gamma radiation generated by the enclosed nuclear or radioactive materials. The system analyzes the detected objects and produces a threat "classification". If a nuclear or radiological threat is detected, a visual and audible alert is presented to the operator as shown in Figure 13.10.

13.6 ScanPyramids Project

Egypt's pyramids withstood centuries of grave robbing after their construction 4500 years ago. In recent times they've been examined and poked at by archaeologists. After thousands of years, it's hard to imagine they have any more secrets.

But amazingly a team of researchers with the ScanPyramids project recently uncovered a previously unknown chamber in the Great Pyramid of Giza. They did it with muon tomography.

ScanPyramids is an Egyptian-International mission designed and led by Cairo University and the French HIP Institute (Heritage Innovation Preservation). This project aims at scanning Old Kingdom Egyptian Pyramids (Khufu, Khafre, the Bent and the Red) to detect the presence of unknown internal voids and structures (Figure 13.11).

ScanPyramids launched in October 2015 and combines several non-invasive and non-destructive techniques which may help to a better understanding of their structure and their construction processes and techniques. The team is currently using Infrared thermography, muon tomography, 3D simulation and reconstruction techniques.

Figure 13.11 Muon telescope setup on Khufu pyramid. (Photo ScanPyramids)

ScanPyramids is an interdisciplinary project mixing art, science and technology. On November 2, 2017, the ScanPyramids team announced its third discovery in the Great Pyramid, a "plane-sized" previously unknown void named the "ScanPyramids Big Void".

13.7 Commercial Muon Systems

Cosmic muons have now become a technological tool and muon imaging a viable technology. Hence, commercialization makes sense. Decision Sciences International Corporation in the USA, a spin-off company from Los Alamos National Laboratory was the first company in this field. Recently, a number of companies employing muon technology have been founded with many of them being spin-offs from research laboratories or universities as shown in Figure 13.12.

Cargo scanning for national security was one of the first applications. This uses the potential of muon tomography to detect special nuclear materials inside cargo containers that are densely packed with other materials so that they cannot be penetrated by X-rays. The field has blossomed to include mining and monitoring of waste containers.

Company	Date Founded	Country	Main Applications
Decision Sciences	2001	USA	Cargo Scanning
Lingacom	2012	Israel	Cargo Scanning
GeoTomography	2013	Canada	Mining Exploration
Muon Systems	2015	Spain	Industrial Scanning
Lynkeos Tech.	2016	UK	Nuclear Safety
Muon Solutions	2016	Finland	Mining Exploration

Figure 13.12 Companies in the field of muon imaging.

Those applications realized thus far may not be the main ones in the future. There may be other civil applications in monitoring structural building stability or using muons for underwater or underground positioning.

13.8 Other Projects Involving Muons

In 2019 M. Abbrescia and colleagues proposed employing the Multi-gap Resistive Plate Chambers (MRPC) of the Extreme Energy Events (EEE) Project (Chapter 11) as muon tracking detectors to monitor the long term stability of civil buildings and structures. The basic idea behind this possibility is to employ a set of position-sensitive detectors fixed to different parts of the structure and reconstruct the muon tracks passing through them. Any misalignment between the positions measured in all detection planes with respect to the original alignment condition could signal a mechanical shift of a part of the structure with respect to other parts.

Variations of this basic strategy would consider a single good tracking muon detector fixed to part of the structure or to the ground, and one or several additional detectors - even without tracking capabilities - fixed to the part of the structure being monitored, and operated in coincidence with the tracking detector. Long lasting measurements of the angular distribution of the cosmic muons passing through the tracking telescope and one of the additional detectors may give information on possible shifts in the relative position of the two devices, thus evidencing long term instability of the structure.

Researchers performed coincidence measurements between a muon tracking detector located in the underground basement and a scintillator based detector placed at various positions on the floors of a building. Early results show that it is possible to observe small shifts of the movable detector with respect to the tracking device, on the order of a few mm.

Another muon application involves underwater and underground positioning techniques. Thus far these techniques have been limited to those using classical waves (sound waves, electromagnetic waves or their combination). However, the positioning accuracy is strongly affected by the conditions of the medium of propagation (temperature, salinity, density, elastic constants, opacity, etc.).

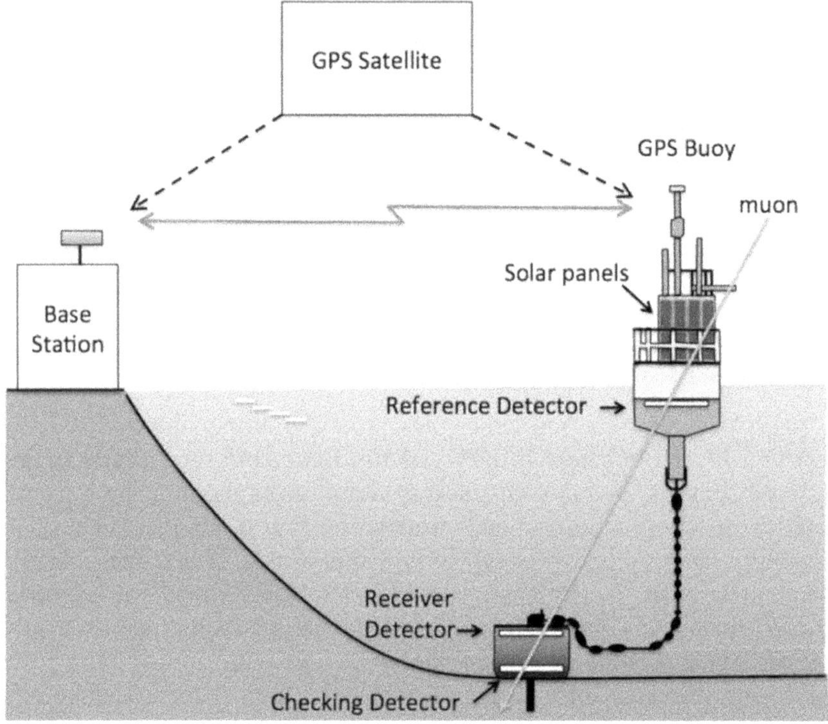

Figure 13.13 Deployment of μPS on a Zosen GPS buoy. (Image HKMT.)

In 2020 Hiroyuki K. M. Tanaka unveiled a Muometric Positioning System (μPS) using cosmic muons as a new underwater and underground positioning technique. It is a precise and entirely new three dimensional method using cosmic muons. This muonic technique is totally unaffected by the media condition and can be universally implemented anywhere on the globe without a signal transmitter. It has the potential to be used for positioning the receiver detector located underwater or underground three dimensionally with a great accuracy within the coordinates defined by the reference detectors. It uses the fact that cosmic muons always precipitate from the upper hemisphere and multiple particle detectors (reference detectors) located above a receiver detector can provide the times of flight between these reference detectors and a receiver detector. This information can be simply converted to the distances between these detectors by multiplying by the speed of light in a vacuum.

Results of the laboratory-based experiments and simulations showed that plate-tectonics-driven seafloor motion and magma-driven seamount deformation can be detected with the μPS. Figure 13.13 shows how to apply the μPS to positioning the seafloor. The ocean-based μPS could be deployed as a combination with the GPS at sea level. With this scheme, the locations of the reference detectors will be determined by the GPS and their positioning accuracy depends on the GPS quality. For example, the reference and receiver detectors could be respectively equipped with the GPS buoys and the anchor. The typical size of the GPS buoy is 5 m in diameter fixed by a 30-ton anchor, having a sufficient capacity to hold particle detectors. In this case, the natural motion of the buoy collects the data for different locations above the receiver detector.

13.9 Measuring Soil Moisture with Cosmic Ray Neutrons

It's hard to believe but neutrons propelled to Earth by cosmic rays from space are helping scientists in more than 25 countries measure water in soil and help farmers save water and adapt to climate change. With a cosmic ray neutron sensor (CRNS), scientists track these fast-moving neutrons in the atmosphere to determine how much water is already in the soil and when the farmer needs to add water to help crops thrive even in harsh climate conditions.

Moisture in the top meter of soil impacts many aspects of human life through its effects on vegetation, surface runoff and land surface interactions with the atmosphere. It's crucial to agriculture, where sufficient water is vital to support germination and plant growth during critical parts of the growing season. Low soil moisture often results in drought with attendant ecological damage. But excess soil moisture is a major factor in the occurrence of surface runoff. Soil moisture is also a major control on energy and water exchanges between the land and atmosphere.

In 2008 Marek Zreda and colleagues (University of Arizona) introduced the idea of measuring soil moisture non-invasively using cosmic ray neutrons. Interactions of cosmic rays with atmospheric nuclei produce almost all neutrons found in the atmosphere and in the top 1 meter of the crust. Neutron intensity in the subsurface is controlled by the scattering and absorption properties of the soil. Hydrogen contained in water dominates the moderating power of a soil. Measurements of cosmic-ray neutrons in variably-saturated soils show that neutron intensity and soil moisture content are inversely correlated.

Figure 13.14 A stationary cosmic ray neutron moisture sensor (CRNS).

A CRNS is capable of estimating soil moisture on a large scale (up to tens of hectares) based on the detection of moderated neutrons in the soil matrix. A typical stationary sensor is shown in Figure 13.14.

Typically, stationary CRNS probes are used to obtain continuous information on field scale soil moisture dynamics. Recently, mobile applications of CRNS probes have been introduced, which enable measuring spatial soil moisture variability at a larger scale. Since 2013, scientists at the Joint FAO/IAEA Division of Nuclear Techniques in Food and Agriculture have been testing and calibrating cosmic ray neutron sensors, including a mobile version in the form of a backpack.

Chapter 14

14 Cosmic Neutrinos

14.1 Introduction

Neutrinos are generated in the Sun and other places in our own galaxy. Some come from far beyond our own galaxy. Travelling through light-years of space, these neutrinos carry information from distant sources directly to our planet.

Cosmic neutrinos are thought to be generated by cosmic rays in powerful, mysterious extragalactic sources that can be thought of as "cosmic accelerators. Figure 14.1 shows one possible accelerator.

Figure 14.1 Artist's illustration of a black hole, one kind of cosmic accelerator that could be sending us high-energy neutrinos. Credit: NASA/CXC/M.Weiss

Cosmic neutrinos are more energetic than any other kind of neutrino we see on Earth, and more powerful than any we can produce in experiments. They're important since they may point to powerful energy sources in the Universe. These sources are a mystery but could include black holes, pulsars, remnants of supernovas, gamma ray bursts, blazars (active galactic nucleus (AGN)), or some other as yet unknown source. Now let's learn more about neutrinos and their history.

14.2 Early History of Neutrinos

There was a measurement puzzle with beta decay in the 1920's which seemed to indicate that the energies of electrons ejected from some radio-active atoms violated the conservation of energy and momentum. To solve this conundrum, Wolfgang Pauli in 1930 postulated a radical new particle - the neutrino - to clarify how beta decay could conserve energy and momentum. He said these undetected neutrinos were also being emitted along with the electrons and they had no electric charge or mass but still carried energy and momentum.

He was greatly bothered by the idea of particles with such outlandish properties. He declared "*I have invented a particle that experimental physicists will never see.*" In 1934 Enrico Fermi took up Pauli's idea and formulated a coherent theory of the neutrino and other particles. Fermi's theory postulates that protons and neutrons interact with the combined field of electrons and neutrinos in such a way that an electron and a neutrino are created and irradiated when a neutron turns into a proton.

Since these early experiments and theoretical interpretations we have grown more sophisticated and now know that there are actually three different types (or flavors) of neutrinos. They are respectively, electron (v_e), muonic (v_u) and tauonic (v_t) each with its own corresponding anti-neutrinos.

Our galaxy is flooded with neutrinos that pass through us without any known effects. Some neutrinos come from the sun, the result of nuclear reactions in the core. Similarly, other stars produce neutrinos but because of their distance we do not notice them unless there is an outburst such as a supernova. Additional neutrinos are produced as the secondary products of cosmic ray collisions at the top of Earth's atmosphere. From these sources, neutrinos constantly flood Earth, at all times and evenly in all directions.

Experiments performed in 1989 at the Stanford Linear accelerator Center (SLAC) and the Large Electron – Positron Collider (LEP) at CERN has shown there only three types of neutrinos. Furthermore they determined that no additional neutrinos beyond the three already known can exist.

14.3 First Neutrino Detection

Neutrinos are subatomic particles similar to an electron, but without electrical charge and a very small mass. They are one of the most abundant particles in the universe. However, since they have very little interaction with matter, they are incredibly difficult to detect. Nuclear forces regard electrons and neutrinos identically; neither participate in the strong nuclear force, but both participate equally in the weak nuclear force.

In the early 1950's, the only supporting evidence of neutrinos was the observation of charged particles as a by-product of various meson decays. Many scientists felt that neutrinos could not be directly detected. But two young physicists, Frederick Reines and Clyde Cowan, working at Los Alamos Scientific Laboratory at the time, thought it might be fun to try.

Figure 14.2 Frederick Reines (left) and Clyde L. Cowan conducted the first experiment to confirm the existence of neutrinos in 1956. (Department of Energy)

In the summer of 1951, Reines and Cowan met by chance on an airplane flight and began discussing what could be done that was interesting in physics. Reines suggested they work on the neutrino. Surprised, Cowan asked, "*Why work on the neutrino*"? To which Reines replied, "*Because everybody said you couldn't detect it*". The two men agreed it would be a fascinating problem. After some reflection, they thought it might be possible to use a reaction that was mentioned in Fermi's theory called inverse beta decay. This involved the interaction of an antineutrino with a proton in a large tank filled with water and some cadmium. The reaction produces a positron and a neutron. The positron slows and annihilates with an electron producing two 0.5 MeV gamma rays which can be detected with large scintillators on the sides of the tank. The neutron is captured by the cadmium also producing detectable gamma rays. They decided to name the project after a ghostly particle, Project Poltergeist.

The scientists used a nuclear reactor, an incredibly dense source of neutrinos, to finally catch the ghost. Reines and Cowan worked first at the Hanford nuclear reactor in the state of Washington. With encouraging results, they moved their experiment to the more powerful Savannah River reactor in South Carolina where they had better shielding against cosmic rays. They used two tanks with a total of about 200 liters of water with about 40 kg of dissolved cadmium chloride. The water tanks were sandwiched between three scintillator layers which contained 110 five-inch photomultiplier tubes. After months of data collection, they had accumulated data at about three neutrinos per hour in their detector. They announced their success in a 1956 telegram to Wolfgang Pauli: "*We are happy to inform you that we have definitively detected neutrinos.*" From experiments such as this as well as neutrinos detected from astronomical sources has come our knowledge of the properties of neutrinos.

14.4 Solar Neutrinos

Originating from the sun, solar neutrinos are the most common type of neutrino, easily passing through a person or any other object on Earth. The sun sends us about 60 billion of these electron neutrinos per square cm per second. They are born from a process known as nuclear fusion that occurs in our sun's core.

A baffling thing happened when scientists started looking for those electron neutrinos in the 1960s. Only about one third of the predicted number of neutrinos actually showed up in detectors. This became known as the solar neutrino problem, and it took nearly four decades to solve it.

The mystery started with the Homestake experiment led by Ray Davis Jr in the late 1960's. It was housed underground in the caverns of the Homestake Gold Mine in South Dakota (Figure 14.3), then an active mine. Because of the expected low neutrino counting rate, this location, 1.5 km underground, was selected to minimize the ever present cosmic ray background. The detector's core is a tank holding 400,000 liters (about 100,000 gallons) of extremely pure perchlorethylene, a cleaning fluid. Solar neutrinos react with some of the fluid's chlorine atoms to create ^{37}Ar, an isotope of argon, which is radioactive with a half-life of 35 days. To detect the argon isotope, Davis flushed the system with helium gas every 100 days to extract the argon. During those 100 days many of the argon atoms decayed, but the radioactivity of the survivors could still be measured.

Davis's colleague, John Bahcall, had predicted how many neutrinos should arrive from the sun and transform one of the chlorine atoms in the detector into an argon atom. But only one third of the neutrinos seemed to arrive. They weren't sure what caused the problem. Was it in Davis's experiment, Bahcall's calculations, the current model of the sun, or their picture of neutrinos? Some scientists, including Bruno Pontecorvo, proposed that the neutrino model was the error, but many were skeptical.

All sorts of explanations were considered in an effort to solve the solar neutrino problem. A number of hypotheses were made including a re-examination of the reactions that take place in the sun. In 1988, John Bahcall and R. K. Ulrich reconsidered the standard solar model. They re-calculated the neutrino flux that could be expected and confirmed a difference between the measured and theoretical values

Figure 14.3 A massive tank, one mile underground in the Homestake Gold mine in South Dakota where Raymond Davis detected solar neutrinos.

Adding to the confusion was the Kamiokande experiment in Japan in 1989. This pure water detector found more neutrinos than Davis's experiment - about half of the predicted number. So, there was still the question of those missing neutrinos. The GALLEX experiment in Italy and SAGE experiment in Russia also found that the expected low-energy neutrinos were missing.

Over time, measurements of the sun improved and the solar model was validated. Researchers then began to look to new physics beyond the Standard Model to explain the neutrino deficit. Two newer experiments, the Super-Kamiokande in 1996 and the Sudbury Neutrino Observatory in

Canada in 1999 provided the breakthrough. Leaders of these projects discovered the solution to the solar neutrino problem: neutrino oscillations. It turned out that roughly two-thirds of the electron neutrinos coming from the sun were changing their flavor as they traveled, arriving as muon or tau neutrinos. This also proved that they have mass, a discovery not predicted by the standard model. Solar neutrinos have now become a quantitative tool for astronomy.

14.5 Neutrino Astronomy

Neutrino astronomy is a relatively new and fast developing branch of astronomy that observes astronomical objects with neutrino detectors in special observatories. As we have learned so far, neutrinos are created in many ways including certain types of radioactive decay, nuclear reactions such as those that take place in the Sun, high energy astrophysical phenomena and when cosmic rays hit atoms in the atmosphere. Neutrinos, unlike photons, rarely interact with matter so that it is unlikely for them to scatter along their path. Thus, neutrinos offer a unique opportunity to observe processes that are inaccessible to optical telescopes, such as reactions in the Sun's core. Neutrinos, unlike cosmic rays, also present a very strong pointing direction.

Because neutrinos interact weakly, they require kilometer scale detectors and large target masses (often thousands of gallons of liquid). The detectors also must employ effective shielding and clever software to remove background signals. Early efforts concentrated on utilizing large volumes of natural water as Cherenkov detectors that see the light produced when neutrinos interact with nuclei in the detector.

In the 1980’s the Deep Underwater Muon And Neutrino Detector Project (DUMAND) was a proposed underwater neutrino telescope to be built in the Pacific Ocean, off the shore of the island of Hawaii, five kilometers beneath the surface. It would have included thousands of strings of instruments occupying a cubic kilometer of the ocean. Work began in 1976, at Keahole Point, but the project was cancelled in 1995 because the technology was not mature enough for this kind of project. Although it was never completed, DUMAND was a precursor of the Antarctic Muon And Neutrino Detector Array (AMANDA), and the water Cherenkov neutrino telescopes in the Mediterranean (ANTARES, NEMO and the NESTOR Project).

Later in the 1980’s, the Antarctic Muon And Neutrino Detection Array (AMANDA) neutrino telescope project was started in the Antarctic ice located beneath the Amundsen–Scott South Pole Station. AMANDA consists of optical modules, each containing one photomultiplier tube, sunk in the Antarctic ice cap at a depth of about 1500 to 1900 meters. It is made up of an array of 677 optical modules mounted on 19 separate strings that are spread out in a circle with a diameter of 200 meters. Each string has several dozen modules, and was put in place by "drilling" a hole in the ice using a hot-water hose, sinking the cable with the attached optical modules, and then letting the ice freeze around it.

AMANDA detects very high energy neutrinos (up to 50 GeV) which pass through the Earth from the northern hemisphere and then react just as they are leaving upwards through the Antarctic ice. Neutrinos interact with nuclei of oxygen or hydrogen atoms contained in the surrounding water ice through the weak nuclear force, producing a muon and a hadronic shower. Cherenkov radiation from these latter particles are detected and analysis of the photon "hits" determines the direction of the original neutrino with a spatial resolution of approximately 2 degrees.

AMANDA's goal was a major step for neutrino astronomy, identifying and characterizing extra-solar sources of neutrinos. Compared to underground detectors like Super-Kamiokande in Japan, AMANDA was capable of looking at higher energy neutrinos because it is not limited in volume to a manmade tank; however, it had much less accuracy because of the less controlled conditions and wider spacing of photo-multipliers. In 2005, after nine years of operation, AMANDA officially became part of its successor project, the IceCube Neutrino Observatory.

14.6 IceCube Neutrino Observatory

AMANDA served as a proof of concept for the IceCube Neutrino Observatory – the first and so far the only, cubic-kilometer neutrino telescope. Francis Louis Halzen (University of Wisconsin-Madison) who had worked on AMANDA argued for this much larger detector. Eventually he secured primary funding for IceCube from the National Science Foundation (NSF) with assistance from partner funding agencies around the world. Halzen is the principal investigator and the University of Wisconsin–Madison is the lead institution, responsible for the maintenance and operations of the detector.

The IceCube Neutrino Observatory is the first detector of its kind, designed to observe the cosmos from deep within the South Pole ice (Figure 14.4). It is a flagship experiment in neutrino and multi-messenger astronomy because of its discovery of very high energy cosmic neutrinos and a blazar. An international group of scientists responsible for the scientific research makes up the IceCube Collaboration.

The Antarctic neutrino observatory, which also includes the surface array IceTop and the dense infill array DeepCore, was designed as a multipurpose experiment. IceCube collaborators address several big questions in physics, like the nature of dark matter and the properties of the neutrino itself. IceCube also observes cosmic rays that interact with the Earth's atmosphere, which have revealed fascinating structures that are not presently understood.

Figure 14.4 The IceCube Neutrino Observatory at the South Pole

Approximately 300 physicists from 53 institutions in 12 countries make up the IceCube Collaboration. The international team is responsible for the scientific program, and many of the collaborators contributed to the design and construction of the detector. Exciting new research conducted by the collaboration is opening a new window for exploring our universe.

IceCube's measuring array consists of spherical optical sensors called Digital Optical Modules (DOMs), each with a photomultiplier tube (PMT) and a single-board data acquisition computer which sends digital data to the counting house on the surface above the array (Figure 14.5). IceCube was completed in December 2010.

DOMs are deployed on strings of 60 modules each at depths between 1,450 and 2,450 meters into holes melted in the ice using a hot water drill. IceCube is designed to look for point sources of neutrinos in the TeV range to explore the highest-energy astrophysical processes.

IceTop consists of 81 stations located on top of the same number of

IceCube strings. Each station has two tanks, each equipped with two downward facing DOMs. IceTop, built as a veto and calibration detector for IceCube, also detects air showers from primary cosmic rays in the 300 TeV to 1 EeV energy range. The surface array measures the cosmic-ray arrival directions in the Southern Hemisphere as well as the flux and composition of cosmic rays. Also see Section 10.11.

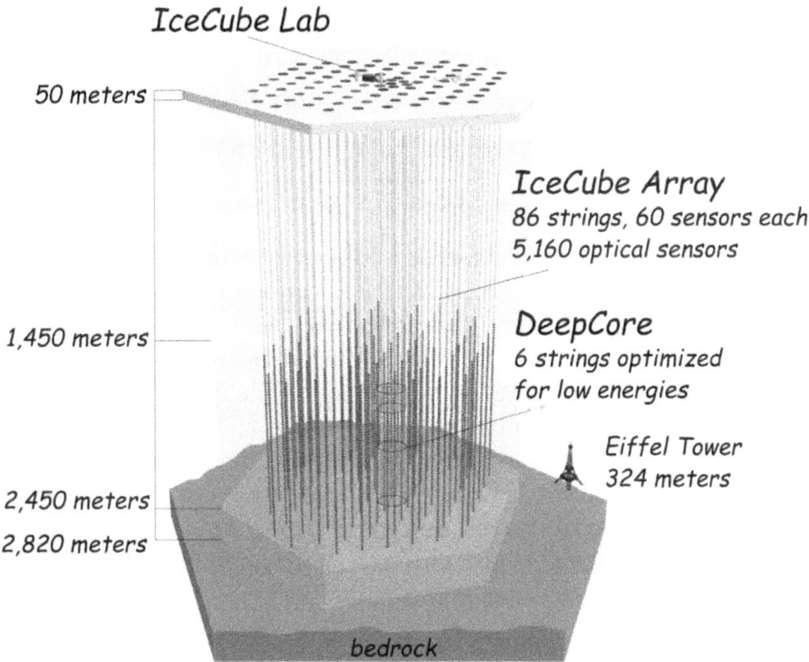

Figure 14.5 The IceCube neutrino observatory array is designed so that 5,160 optical sensors view a cubic kilometer of clear South Polar ice.

Developments in neutrino astronomy have been driven by the search for the sources of cosmic rays, leading at an early stage to the concept of a cubic-kilometer neutrino detector. In November 2013 it was announced that IceCube had detected 28 neutrinos that likely originated outside the Solar System.

14.7 IceCube Neutrino Observatory Hits the Jackpot

On September 22, 2017 a ghostly neutrino that has been hurtling unhindered for 3.9 billion years through the cosmic vastness of space was detected in the deep recesses of the south polar ice cap. It made its presence known by annihilating itself in a collision with an atom in the frozen darkness 2 kilometers beneath the surface.

The crystal-clear ice beneath the South Pole provides the medium that allows the Nation Science Foundation's (NSF) IceCube to document the interaction of neutrinos with terrestrial matter. Collisions between high-energy neutrinos and atomic nuclei are very rare but produce an unmistakable signature - a characteristic cone of blue light that is mapped through the detector's grid of 5,000 photomultiplier tubes.

When a neutrino slams into the nucleus of an atom, it creates one or more secondary charged particles, which, in turn, create the blue light. Because the charged particle and light it creates stay essentially true to the neutrino's direction, it gives scientists a path to follow back to a source.

Data on this amazing collision was gathered by IceCube's Neutrino Observatory at the Amundsen-Scott South Pole Station in Antarctica. It likely points to an answer of a more than century-old riddle about the origins of high-energy cosmic rays.

This subatomic particle's demise - labeled as neutrino event 170922A - triggered a worldwide cascade of astronomical observations using a host of varied technologies. That the detection was confirmed by other instruments, including an orbiting telescope operated by NASA, is a demonstration of the value of the emerging field of "multi-messenger astronomy," This is the ability to quickly organize instruments globally to verify discoveries by combining data from messenger signals. NSF Director France Córdova said that "Each messenger - from electromagnetic radiation, gravitational waves and now neutrinos - gives us a more complete understanding of the universe, and important new insights into the most powerful objects and events in the sky".

Since cosmic rays were first detected in 1912 they have posed an enduring mystery. How are they created and what propels them across cosmic distances before they collide with Earth? Where do they originate?

Trying to trace these charged cosmic rays back to their points of origin is extremely difficult as the magnetic fields that permeate space alter their trajectories. But the powerful, naturally occurring cosmic accelerators that produce cosmic rays also produce cosmic neutrinos. Neutrinos are uncharged particles, unaffected by even the most powerful magnetic fields. Because they rarely interact with matter and have almost no mass, neutrinos travel nearly undisturbed, giving scientists an almost direct pointer to their source.

The International researchers that made this discovery traced the path of this neutrino detected to a previously known blazar, designated as TXS 0506+056. This is a little studied blazar, the nucleus of a giant galaxy that fires off particles in massive jets of elementary particles, powered by a supermassive black hole at its core. Francis Halzen (University of Wisconsin-Madison) and lead scientist for the IceCube Neutrino Observatory said, "The evidence for the observation of the first known source of high-energy neutrinos and cosmic rays is compelling,"

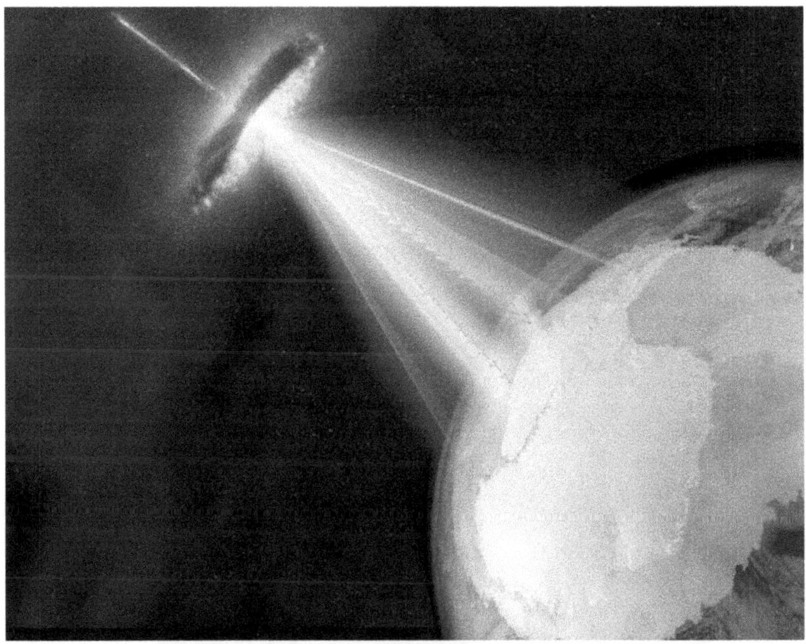

Figure 14.6 In this artistic rendering, a blazar emits both neutrinos and gamma rays that could be detected by telescopes on Earth and in space. Credit: IceCube NASA

IceCube's alert system - triggered when neutrinos of very high energies crash into an atomic nucleus in or near the IceCube detector - sent coordinates to telescopes worldwide less than a minute after detection for follow-up observations. Two gamma ray telescopes looked in the direction provided by IceCube and picked up signals. They were NASA's orbiting Fermi Gamma-ray Space Telescope and the Major Atmospheric Gamma Imaging Cherenkov Telescope (MAGIC) in the Canary Islands. They detected a flare of high-energy gamma rays associated with TXS 0506+056. This convergence of observations helped identify the blazar as the source. See target in Orion constellation of Figure 14.7.

Fermi was the first telescope to identify enhanced gamma-ray activity from TXS 0506+056 within 0.06 degrees of the IceCube neutrino direction. Over a decade of Fermi observations of this source, this was the strongest flare in gamma rays, the highest-energy photons. A later follow-up by MAGIC detected gamma rays of even higher energies.

High-energy gamma rays can be produced either by accelerated electrons or protons. The observation of a neutrino, a hallmark of proton interactions, is the first definitive evidence of proton acceleration by black holes. Halzen said, "Now, we have identified at least one source of cosmic rays because it produces cosmic neutrinos,"

Figure 14.7 The constellation of Orion, with a bullseye on the location of the blazar. Silvia Bravo Gallart/ Project_WIPAC_Communications, CC BY-ND

Following this neutrino detection, the IceCube team quickly searched the detector's archival data. They discovered a flare of neutrinos from December 2014, coincident with the same blazar, TXS 0506+056, which scientists have nicknamed "the Texas source."

These observations show that TXS 056+056 is among the most luminous sources in the known universe. It adds to a growing body of data that indicates that the blazar is the first known source of high-energy neutrinos and high-energy cosmic rays.

There are a number of observatories in space and here on Earth that have collaborated in this discovery. Their missions include gamma rays, x-rays, optical, radio and neutrino observations:

Gamma rays: Space missions, AGILE gamma ray telescope, Integral space telescope, and Fermi gamma-ray space telescope. Ground based telescopes, HAWC in Mexico, H.E.S.S. in Namibia, MAGIC in Spain, and VERITAS in the U.S.

X-rays, optical and radio radiation: Space missions MAXI, NuSTAR, and Swift. Ground based observatories, ASAA-SN in Chile and the U.S., GTC in Spain, Kanata in Japan and the U.S., Kiso astronomical observatory in Japan, Liverpool telescope in Spain, ORVO in the U.S., SALT in South Africa, Suburu in Japan, and VLA in the U.S.

Neutrinos by ANTARES in France.

International teams run these observatories with over a thousand scientists supported in funding agencies around the world.

Chapter 15

15 Too Interesting To Be Left Out

15.1 Introduction

During the writing of this book many interesting items related to cosmic rays were found that did not seem to fit in any particular category. These items include: real time cosmic ray monitoring stations, using the Hubble telescope as a cosmic ray detector and a plan to make the moon an enormous cosmic ray detector. Please use the web links provided to see more information.

15.2 NOvA as a Cosmic Ray Detector

Fermilab's NOνA experiment is helping scientists determine the role that ghostly particles called neutrinos played in the evolution of the cosmos.

NOνA scientists use a 300-ton particle detector at Fermilab (the near detector) and a 14,000-ton detector in northern Minnesota (the far detector) to study neutrino oscillations. The near detector sits in a cavern 350 feet underground and measures the composition of the neutrino beam as it leaves the Fermilab site. The neutrinos oscillate as they travel straight through the Earth. The far detector records the types of neutrinos that arrive in Minnesota.

Fermilab's accelerator complex produces the most intense high-energy neutrino beam in the world and sends it straight through Earth's crust to northern Minnesota, no tunnel required. Moving at close to the speed of light, the neutrinos make the 500-mile journey in less than three milliseconds. Neutrinos rarely interact with matter. When a neutrino smashes into an atom in the NOνA detector in Minnesota, the resulting particles create distinctive particle tracks. Scientists explore these particle interactions to better understand the transition of muon neutrinos into electron neutrinos. The experiment also helps answer important scientific questions about neutrino masses, neutrino oscillations and the role neutrinos played in the early universe.

Fermilab provides real time event displays from both the NOvA *Far* detector and the NOvA *Near* detector. While these displays were designed for neutrinos they also provide a fascinating look at cosmic ray tracks. Sometimes, if you are lucky, you can see an actual cosmic ray shower. To access these live displays go to: https://nusoft.fnal.gov/nova/public/

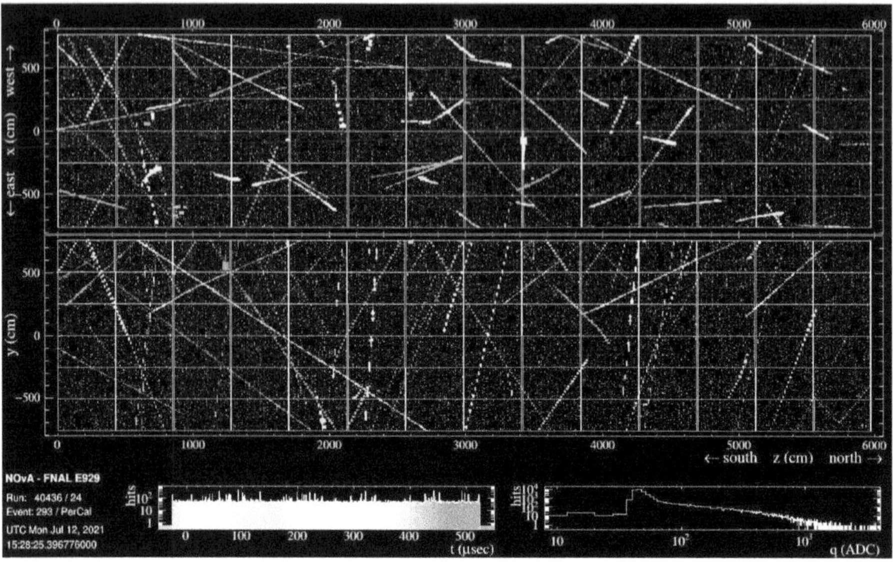

Figure 15.1 Event display from NOvA Far detector. This image is normally live data, which refreshes every 15 seconds. The top large rectangle is the view from above; the bottom is the view from the side. The NuMI beam from Fermilab is coming in from the left of the picture. During a 500 us long capture window the display shows mainly long straight tracks from cosmic ray muons.

15.3 Neutron Cosmic Ray Monitors

The Bartol Research Institute (University of Delaware) neutron monitor web site has 8 on-line monitoring stations with links to other monitoring stations. A neutron monitor is an instrument that measures neutrons produced from high-energy particles impacting Earth from space. Because the intensity of cosmic rays hitting Earth is not uniform, it is important to place neutron monitors at multiple locations in order to form a complete picture of cosmic rays in space.

At the Bartol site, you can obtain information about their observing stations and learn why scientists study cosmic rays. You can view real-time data from the neutron monitors, and you can download past neutron monitor data. They can be found at: http://neutronm.bartol.udel.edu/.

An example of cosmic ray neutron data from the Inuvik neutron monitoring station with ten minute averaging is shown in Figure 15.2.

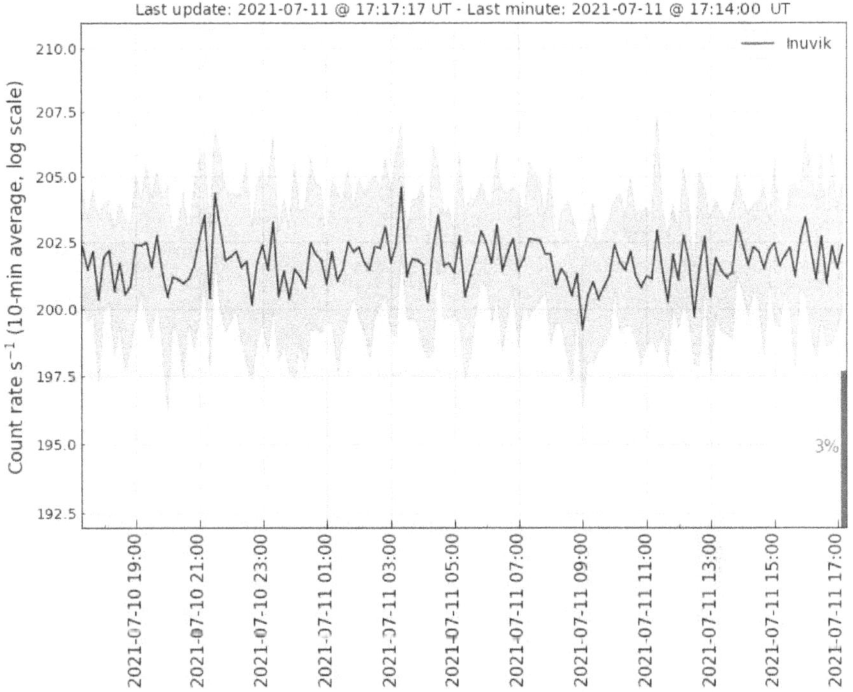

Figure 15.2 Data from Bartol Research Institute real time neutron monitoring station at Inuvik, a town in the Northwest Territories of Canada.

15.4 Hubble Telescope as a Cosmic Ray Detector

The Hubble Space Telescope (HST) has been operational for over 30 years and throughout that time it has been bombarded by high energy cosmic rays. But for most of this time, astronomers have considered them a nuisance or a nightmare. Neither ground nor space telescopes are immune to the barrage of the energetic particles. When a particle travels through a camera, it leaves a sharp and bright trace on the resulting image. Obviously, astronomers try to remove this pollution from their data. But now scientists have shown that this trash may actually be useful.

Using the Hubble Space Telescope as a particle detector, astronomers have traced cosmic rays flowing in Earth's geomagnetic field. A team of scientists, in a pilot study, have demonstrated how to use this particle data (Figure 15.3). Their work appears in *The Astrophysical Journal*, with preprint available here: https://arxiv.org/pdf/2006.00909.pdf.

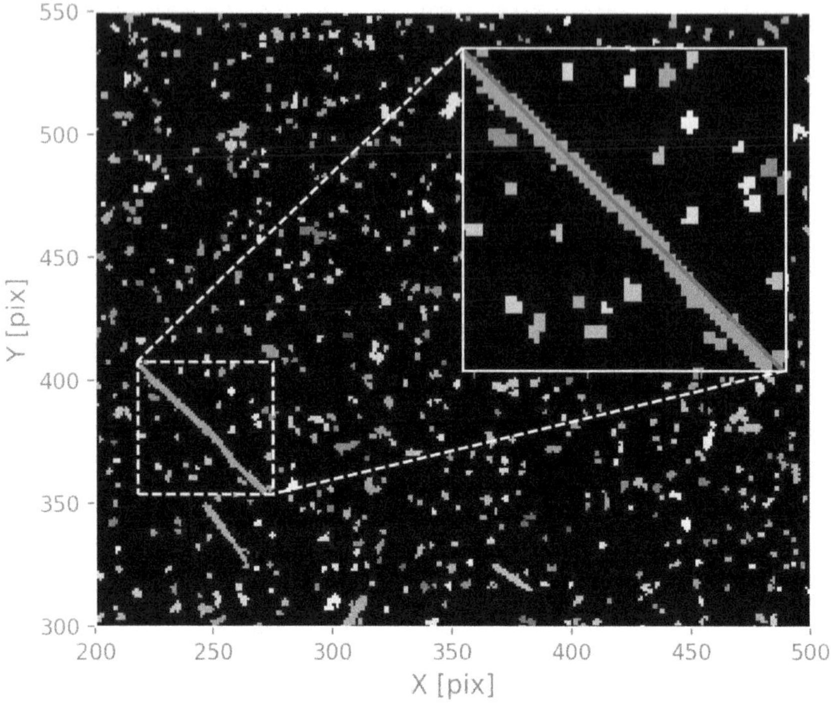

Figure 15.3 Hubble cosmic ray image. It shows the projected path computed for a single, elongated cosmic ray in a reconstructed CCD segmentation map.

Susana Deustua (Ars Metrologia) who led the study said the basic idea for the project came from an international group of astronomers, particle physicists, and planetary scientists. She said, "We know that Hubble collects charged particles on its detector, therefore we should be able to glean information about the field from Hubble."

Hubble's rich data archive was the source of images. They found calibration images that were well-suited for their plan and ended up with almost 100,000 images collected over the past 25 years. They used algorithms for finding and removing cosmic ray traces as well as determining how much energy the particle lost in the process.

One of the authors, graduate student Nathan Miles (University of California) developed software to extract such information to carry out the time-demanding computations. His algorithm harvested more than 1 billion cosmic rays from the images. Results provide an important proof of concept with encouraging results. Cosmic ray properties from the Hubble data match those from the PAMELA experiment, an obsolete particle detector in low Earth orbit. They also saw the South Atlantic Anomaly, the famous dip in the Earth's magnetic field and the response of cosmic rays to the solar cycle. So far, the analysis has only scratched the surface. The team hopes to process more of the data to better understand the relation between galactic cosmic rays, the Sun, and Earth's environment.

15.5 Using the Moon as a Cosmic Ray Detector

Scientists from the University of Southampton want to turn the Moon into a giant particle detector to help understand the origin of Ultra-High-Energy (UHE) cosmic rays -- the most energetic particles in the Universe.

Little is known about where these extremely rare cosmic rays come from or how they get their enormous energies. Scientists detect them on Earth at a rate of less than one particle per square kilometer per century.

Justin Bray (University of Southampton) and colleagues have proposed to use the Square Kilometer Array (SKA) to detect vastly more UHE cosmic rays by using the Moon as a giant cosmic ray detector.

The SKA project is an international effort to build the world's largest radio telescope, with eventually over a square kilometer (one million square meters) of collecting area (Figure 15.4). This represents a huge leap forward in both engineering and research towards building and delivering a unique instrument. Detailed design and preparation are now well under way. As one of the largest scientific endeavors in history, the SKA will bring together a wealth of the world's finest scientists, engineers and policy makers to bring the project to fruition.

The SKA will eventually use thousands of dishes and up to a million low-frequency antennas that will enable astronomers to monitor the sky in unprecedented detail and survey the entire sky much faster than any system currently in existence.

On Earth, physicists usually detect high-energy cosmic rays when they hit the upper atmosphere triggering a cascade of secondary particles that generate a short and faint burst of radio waves only a few nanoseconds long. It is this signal that astronomers hope to pick up from the Moon, but as these signals are so short and faint no radio telescope on Earth is currently capable of picking them up.

With its large collecting area and high sensitivity, the SKA will be able to detect these signals using the visible lunar surface -- millions of square kilometers -- giving the researchers access to more data about UHE cosmic rays than they have ever had before.

In a recent paper: https://arxiv.org/pdf/1408.6069v1.pdf a group of astronomers disclosed how the entire moon can be turned into an instrument for studying cosmic rays. That's because upon impact, cosmic rays generate a cascade of secondary particles, which can be seen as nanosecond burst of radio waves. By analyzing these radio waves, scientists might figure out where the rays are coming from.

Figure 15.4 Part of the Square Kilometer Array (SKA) project, an international effort to build the world's largest radio telescope

This effect is complicated by the fact that radio pulses are projected forward in a cone and cannot travel far through the lunar surface before being absorbed. That means that astronomers will only be able to see the radio pulses from ultrahigh energy cosmic rays that graze the edge of the Moon coming our way. So the instrument they need to detect the signal is a highly sensitive radio telescope on Earth, like the SKA.

Appendix 1. Distances in the Universe

Space beyond Earth is so incredibly vast that units of measure which are convenient for us in our everyday lives can become enormous. Distances in the solar system and especially between the stars, grow to be so big when expressed in miles and kilometers that they're unwieldy. So for cosmic distances, we use other types of units: *astronomical units*, *light years* and *parsecs*.

Astronomical Units: Distances in the solar system are often measured in astronomical units (AU). An AU is the average distance between the Earth and the Sun: 1 AU = 1.496×10^8 km = 93 million miles.

Jupiter is about 5.2 AU from the Sun and Pluto is about 39.5 AU from the Sun. The distance from the Sun to the center of the Milky Way is approximately 1.7×10^9 AU.

Light Years: Distances between stars is often measured in light years (ly). A light-year is the distance that light travels in a vacuum in one year: 1 ly = 9.5×10^{12} km = 63,240 AU.

Proxima Centauri is the nearest star to Earth (other than the Sun) and is 4.2 light-years away. This means light from Proxima Centauri takes 4.2 years to travel to Earth.

Parsecs: Many astronomers use parsecs (pc) to measure distance to stars. This unit is closely related to a method of measuring the distances between stars. A parsec is the distance at which 1 AU subtends an angle of 1 arcsec. 1 pc = 3.09×10^{13} km = 3.26 ly

Appendix 2. Powers of Ten Notation

In this book, the distances and sizes of the objects we mention vary from very small, including atoms and atomic nuclei, to very large including galaxies, clusters of galaxies and the size of the universe. To describe such a huge range, we need a way to avoid confusing terms like "a billion trillion" and "a millionth".

Scientists use a system called powers-of-ten notation, which consolidates all of the zeros that you would normally find attached to very large or small numbers such as 1,000,000,000 or 0.0000001. All of the zeros are put in an exponent, which is written as a superscript, and indicates how many zeros you would need to write out the long form of the number. For example:

$10^0 = 1$

$10^1 = 10$
$10^2 = 100$
$10^3 = 1000$
$10^6 = 1,000,000 = 1$ million
$10^9 = 1,000,000,000 = 1$ billion

In powers-of-ten notation, numbers are written as a figure between one and ten multiplied by a power of ten. So for example, the distance to the Moon of 384,000 km can be re-written as 3.84×10^5 km.

Very small numbers can also be written using powers-of-ten notation. The exponent is negative for numbers less than one and indicates dividing by that number of tens. For example:

$10^0 = 1$

$10^{-1} = {}^1/_{10} = 0.1$

$10^{-2} = {}^1/_{10} \times {}^1/_{10} = 0.01$

$10^{-3} = {}^1/_{10} \times {}^1/_{10} \times {}^1/_{10} = 0.001$

$10^{-6} = 0.0000001 = 1$ millionth

$10^{-9} = 0.0000000001 = 1$ billionth

Once again, numbers are written as a figure between one and ten multiplied by a power of ten. So for example, a number like 0.00000375 would be expressed as 3.75×10^{-6}.

Appendix 3. The Electron Volt

The electron-volt (eV) is the standard unit of energy in science. One electron-volt is the energy gained by an electron that is being accelerated by an electric potential difference of 1 volt. One eV is equivalent to $1.602176 \cdot 10^{-19}$ Joule. One Joule is the energy equal to the force of one Newton acting through one meter.

Multiples of eV that are commonly used are:

kilo-electron-volt: 1 keV = 1000 eV
Mega-electron-volt: 1 MeV = 1,000,000 eV $=10^6$ eV
Giga-electron-volt: 1 GeV = 1,000,000,000 eV $=10^9$ eV
Tera-electron-volt: 1 TeV = 1,000,000,000,000 eV $=10^{12}$ eV.

The energy that is necessary to remove an electron from an atom is typically in the range of between a few and a few dozen eV. Typical energies of x-ray photons are in the keV range. The mass of an electron is 511 keV, that of a proton 938 MeV. Each proton in the proton beams of the

Large Hadron Collider, the particle accelerator at the CERN laboratory, is accelerated to an energy of about 7 TeV.

Appendix 4. Energy, Force and Momentum

Potential energy is energy stored in an object or system of objects. To give a few examples, it can be related to the position of the objects, the bonds in a chemical structure, an atom's potential for radioactive decay or even an object's shape. . It has the ability or potential to be transformed into more obvious forms like kinetic energy. Potential energy and kinetic energy are what make up mechanical energy.

Imagine lifting a ball off of the ground and holding it in your hand. Energy is required to lift the ball, but after the ball stops moving, the energy used to lift it does not disappear. It has only been converted into gravitational potential energy. Dropping the ball will release the stored energy; as the ball falls, it transforms the potential to a kinetic form as it moves. When the ball hits the ground, the kinetic energy will be absorbed by the ground as heat or released as sound that you can hear.

Likewise if a toy Slinky is stretched and let go, it will pull itself back to its original shape. That means that there's potential for energy in the Slinky's shape. This is another form of potential energy called elastic potential energy. Some types of potential energy are:

- Gravitational potential energy
- Chemical energy
- Nuclear energy
- Elastic potential energy, also called spring energy
- Electrical potential energy especially in a capacitor

Kinetic energy is the energy of motion. This can be the motion of large objects (macroscopic kinetic energy), or the movement of small atoms and molecules (microscopic kinetic energy). Macroscopic kinetic energy is "high quality" energy, while microscopic kinetic energy is more disordered and "low-quality." In general for many systems the energy flows back and forth between kinetic energy and gravitational potential energy. Friction can also cause macroscopic kinetic energy to become microscopic kinetic energy. Rotational kinetic energy is also a form of kinetic energy that comes from an object spinning.

Macroscopic kinetic energy is the most obvious form of energy as it is the easiest to observe. This is the energy possessed by moving objects. The larger an object is or the faster it moves, the more kinetic energy it has. The sum of potential energy and macroscopic kinetic energy is called mechanical energy and stays constant for a system when there are only conservative forces (no non-conservative forces).

Kinetic energy is calculated using the following formula: $E=\frac{1}{2}mv^2$

- E is energy, measured in joules (J)
- m is mass, measured in kilograms (kg)
- v is velocity, measured in meters per second (m/s)

The more mass a moving object has, the more kinetic energy it will possess at the same speed. A 2000 kg car moving at 14 m/s has twice as much kinetic energy as a 1000 kg car moving at an equivalent 14 m/s.

Because the velocity term in this formula is squared, velocity has a much larger effect than mass does on kinetic energy. A car moving at twice the speed of another car of identical mass will have 2^2 or four times as much kinetic energy. A car moving at three times the base speed will have 3^2 or nine times the original kinetic energy!

Some ways to harness macroscopic kinetic energy include:

Wind power harnesses the kinetic energy possessed by moving bodies of air (wind), converting it into electricity. Wind itself is created initially through complex patterns of changes in thermal energy as the atmosphere and oceans are heated by the sun or cooled in its absence.

Hydropower harnesses the kinetic energy of moving water as it falls (in a waterfall or hydroelectric dam).

Tidal power harnesses the energy of moving water as it moves back and forth due to tides.

Forces are interactions between objects - a push or pull. Forces have the ability to make an object speed up, slow down, change direction or change its shape. Force is conventionally measured in units of newtons (N) or pounds (lbs).

Exerting a force over some distance represents a specific transfer of energy also known as work. Various fields (electric, magnetic, gravitational) are spatial regions that exert forces on objects. The positions of objects within these fields determine an object's potential energy. An object that moves along the push of these fields loses potential energy and gains kinetic energy (or thermal energy). Forces exchanging energy between potential and kinetic energy is an example of mechanical energy conservation.

Momentum is a property of an object's motion. When a push or a pull (a force) acts on an object and changes its motion, the quantity that gets changed is momentum. Energy is required to change the magnitude (size) of momentum, but not its direction. Specifically, momentum (written as *p*) is the mass (m) of the object times its velocity (or speed with direction):

$$\vec{p} = m\vec{v}$$

Massive objects have more momentum for a given speed, while lighter objects have less momentum. This is why it takes more effort (force) to stop a fully loaded truck than an empty one. Likewise, faster moving objects have more momentum than slower moving objects.

Appendix 5. Mass Per Unit Area

Cosmic rays are progressively absorbed as they pass through material. When comparing radiation absorbers of different substances it becomes necessary to consider the density (mass per unit volume) as well as the thickness of the absorbers. Obviously, 1 meter of air will absorb must less radiation than 1 meter of water. This is because the density of air (0.00129 g/cm^3) is much smaller than that of water (1.0 g/cm^3) at standard temperature and pressure.

Scientists find it convenient to define an absorber not by its geometrical thickness but by the mass of a cylinder of unit cross sectional area. We measure the matter passed through by the mass of the material in this cylinder. Its units are g/cm^2.

For example, a layer of water 100 cm thick has a mass per unit area of 100 g/cm^2. While a 100 cm layer of air has a mass per unit area of 0.129 g/cm^2. The atmosphere of Earth has a mass per unit area of about 1000 g/cm^2 and is known as the *atmospheric depth*.

Appendix 6. Particle Physics

Particle physics (also known as **high energy physics**) is a branch of physics that studies the nature of the particles that constitute matter and radiation. Although the word *particle* can refer to various types of very small objects (e.g. protons, gas particles, or even household dust), *particle physics* usually investigates the irreducibly smallest detectable particles and the fundamental interactions necessary to explain their behavior.

In current understanding, these elementary particles are excitations of the quantum fields that also govern their interactions. The currently dominant theory explaining these fundamental particles and fields, along with their dynamics, is called the Standard Model. Thus, modern particle physics generally investigates the Standard Model and its various possible extensions, e.g. to the newest "known" particle, the Higgs boson, or even to the oldest known force field, gravity.

Appendix 7. Inverse Square Law.

This is an illustration of the inverse square law. Here S represents the light source, while r represents the measured points.

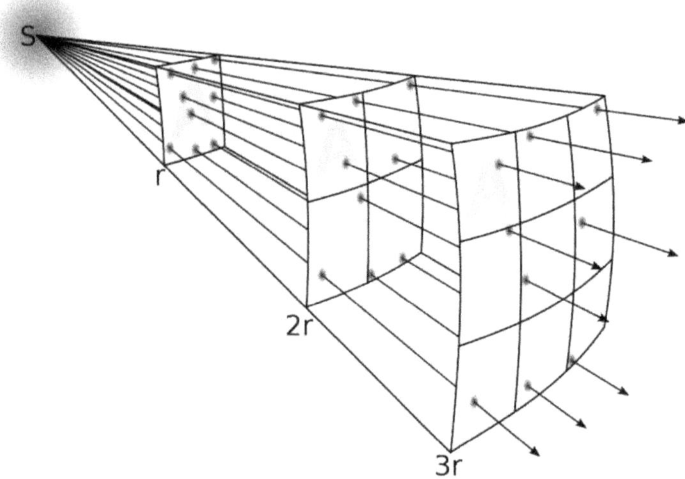

The lines represent the flux emanating from the source. The total number of flux lines depends on the strength of the light source and is constant with increasing distance. A greater density of flux lines (lines per unit area) means a stronger energy field. The density of flux lines is inversely proportional to the square of the distance from the source because the surface area of a sphere increases with the square of the radius. Thus the field intensity is inversely proportional to the square of the distance from the source.

Notable References

Most of the important scientific references on cosmic rays can be found in the physics and astronomy journals or web sites. There are, however, a number of notable books and monographs on cosmic rays that are worth mentioning from a historical or technical viewpoint.

One of my favorites is *Cosmic Rays Thus Far* by Harvey Lemon published in 1936. It has a wonderful foreword by Arthur Holley Compton and some interesting speculations about the nature of cosmic rays. Therein is a happy characterization of the subject by Dr. K. K. Darrow. To wit: "*Unique in the annals of modern physics for the minuteness of the phenomena, the delicacy of the instruments, the adventurous excursions of the observer, the intricacy of the arguments, and the grandeur of the inferences, is the subject of cosmic rays.*"

Another one is *What Are Cosmic Rays?* by Pierre Auger published in 1945. It was originally published as *Rayons cosmiques* in Paris in 1941. The work was completely revised and translated to English by Maurice M. Shapiro. It was written primarily for the reader who lacks technical knowledge in physics but who wants to keep in touch with current developments in science. It has some wonderful photographs of various places measuring cosmic rays, such as the Jungfraujoch in the Swiss Alps or of the laboratory at the top of Mount Evans in Colorado. There are also some excellent cloud chamber photographs including some chambers in magnetic fields.

W. F. G. Swann has written several monographs such as *The Nature of Cosmic Rays* in 1945 and the *Story of Cosmic Rays* in 1957 for the Smithsonian Institution. Included are some nice photographs of early Geiger Mueller counter equipment used in the experiments at the Bartol Research Foundation and around the world.

Perhaps the most comprehensive book, up to this point, is *Cosmic Rays* by L. Janossy written in 1950. He presents a nice historical introduction followed by a very technical presentation on experimental technique and the nature of cosmic rays. His survey of this complex subject includes all the necessary mathematics and is not for beginners.

One of the most widely read classics is *Cosmic Rays* by Bruno Rossi published in 1964. He wrote the book "*for any person with a curiosity about science.*" He presents the material in a nice style and keeps the mathematics to a minimum. Rossi was one of the cosmic ray pioneers and made significant contributions to the field from 1924 onwards.

Martin A. Pomerantz's *Cosmic Rays* in 1971 provides an eloquent survey of the field up to that date. It is more technical than most of those mentioned above and concerns his research interests of terrestrial and solar aspects of cosmic rays.

A. M. Hillas's book *Cosmic Rays* in 1972 is an excellent introduction to the subject by a technical reader in cosmic physics at the University of Leeds. Part 1 is a nice treatment of cosmic rays from Victor Hess up to 1972. Part 2 is a collection of cosmic ray papers by famous authors such as Hess, Bothe and Kolhorster, Carl Anderson, J. R. Oppenheimer, Pierre Auger, W.F.G. Swann, Enrico Fermi among others.

Very interesting is *The Early History of Cosmic Rays* by Y. Sekido, and H. Elliot offered in 1985. This is a 431 page personal recollection of many researchers like Carl Anderson, Pierre Auger, Scott Forbush, George D. Rochester Yataro Sekido and others. It contains many photographs and other artifacts from the period.

An excellent technical book published in 1990 is *Cosmic Rays and Particle Physics* by Thomas K. Gaisser. It concentrates on the highest energy cosmic rays and where they may originate, acquire energy and interact.

A more recent book is *Cosmic Bullets* by Roger Clay and Bruce Dawson published in 1997. It is a popular treatment of cosmic rays with an easy to understand introduction to the subject. But in the later sections it focuses on their own research area of high energy particles.

Michael W. Friedlander's *A Thin Cosmic Rain* in 2000 is a stylish account of the story of cosmic rays. He covers the subject from the early days up to the cosmic neutrinos. His presentation is interesting but may be hard to follow at times when he wanders off into other areas of physics.

Written in 2013 is *Celestial Messengers* by Mario Bertolotti. It is not a popular treatment but a factual book with extensive references. It has some errors, possible due to translations of documents, so it may be disordered in places.

Strictly for historians is the 2014 book *Cosmic Ray History* by Lev Dorman and Irina Dorman. This 784 page book covers the epochs with many details. But it is not for general readers as the English translation and the extensive referencing makes it difficult to follow.

Glossary

Active galaxy. Galaxies that have a small core of emission embedded at the center of an otherwise typical galaxy. This core is typically highly variable and very bright compared to the rest of the galaxy.

Atmosphere. A layer of gases surrounding a planet, moon, or star. The Earth's atmosphere is 120 miles thick and is composed mainly of nitrogen, oxygen, carbon dioxide, and a few other trace gases.

Aurora. A glow in a planet's ionosphere caused by the interaction between the planet's magnetic field and charged particles from the Sun. This phenomenon is known as the Aurora Borealis in the Earth's northern hemisphere and the Aurora Australis in the Earth's Southern Hemisphere.

Black hole. A place in space where gravity is so strong that even light cannot get out. The gravity is so strong because matter has been squeezed into a tiny space.

Cherenkov radiation. This is light produced by charged particles when they pass through an optically transparent medium at speeds greater than the speed of light in that medium. Devices sensitive to this particular form of radiation, called Cherenkov detectors, have been used extensively to detect the presence of charged subatomic particles moving at high velocities. Cherenkov radiation, when it is intense, appears as a weak bluish white glow in the pools of water shielding some nuclear reactors.

Cold Dark Matter. An exotic type of matter that does not interact with light. It is transparent, or "dark," and astronomers only know of its existence by its gravitational influence on ordinary matter.

Cosmic microwave background (CMB). Radiation that is the remnant heat left over from the Big Bang - its afterglow. CMB radiation was emitted long before stars or galaxies ever existed. It blankets the universe today at a rather chilly and uniform 2.73 Kelvin. Temperature fluctuations in the CMB today reflect density fluctuations moments after the Big Bang. These density fluctuations gave rise to the vast web-like structure of galaxies and voids we see today.

Cosmic rays. Extremely energetic atomic nuclei (including protons) from space that bombard the Earth from all directions.

Electromagnetic cascade. A flow of electrons, positrons, and gamma rays in the atmosphere initiated by a high energy gamma ray. Extensive air showers initiated by a cosmic ray particle contains many electromagnetic cascades, each initiated by a gamma ray from a pion decay.

Electromagnetic radiation. Includes many types of radiation such as gamma rays, X-rays, ultraviolet light, visible light, infr-red radiation, microwaves and radio waves.

Energy spectrum. A plot of the range of energies present in a sample of cosmic rays together with the number of particles present at each energy.

Extensive air shower. A flow of subatomic particles in the Earth's atmosphere initiated when an energetic cosmic ray collides with an atmospheric atom or molecule.

Flourescent light. Light produced through the excitation and ionization of atoms and molecules by high speed charged subatomic particles. This light is emitted in all directions unlike beamed Cherenkov light.

Galaxies. Immense aggregations of stars, gas, and dust held together by their mutual gravity. Our home, the Milky Way galaxy, is a typical large, spiral-shaped galaxy. It is about 100,000 light years across, and our Sun is one of about 100 billion stars that populate the Milky Way galaxy.

Gamma rays. The most energetic form of electromagnetic radiation.

Ionization. A process whereby an atom loses electrons. It may occur when a high speed charged particle passes through a collection of atoms.

Isotope. A form of chemical element whose nuclei have the same number of protons but a different number of neutrons.

Muon. A subatomic particle of the lepton family, a sister particle to the electron. It has a mass of about 200 times the electron and exists in both positive and negative types.

Neutron. A subatomic particle with no charge, mass slightly greater than that of a proton, and spin of ½.

Neutron star. An extremely compact star composed almost entirely of neutrons at a density equal to that of an atomic nucleus.

Pion. A subatomic particle of that exists in three varieties (neutral, positive charge or negative charge) with a mass of about 250 times that of an electron.

Secondary cosmic ray. A high speed subatomic particle that exists as part of an extensive air shower started by a primary cosmic ray.

Solar Flare. A bright eruption of hot gas in the Sun's photosphere.

Supernova. An explosion at the end of the life of a massive star which may leave behind the stellar core in the form of a neutron star or black hole.

Index

absorption curves, 37
Abundance of nucleons, 159
Akeno Giant Air Shower Array, 193
Alpha Magnetic Spectrometer, 168
Alpha particles, 37
Alvarez, 82, 223
Anderson, 93
Arrowhead Lake, 27
atomic structure, 32
Auger, 169, 172, 173, 174, 175

Bartol Research Foundation, 54
Becquerel, 9, 88
Beta particles, 38
Bethe and Heitler, 103
Blackett, 60
Blau and Wambacher, 131, 140
Bohr, 34, 90
Bowen, 24
bremsstrahlung, 39

Cameron, 27
Chadwick, 34, 91
Chicago Air Shower Array, 194
Clay, 68
cloud chamber, 40
coincidence technique, 55
Compton, 50, 69, 75, 117
Compton scattering, 44
controlled cloud chamber, 60
Conversi, 126
Cosmic neutrinos, 239
cosmic ray telescope, 81, 85
Cosmic@Web, 209
Coulomb, 7
CREDO, 207
Curie, 9

Davis, 242
Decision Sciences, 228, 231
Dirac, 48, 97
Direct detection method, 162

Earth's magnetic field, 63
Earth's *magnetosphere*, 69
east-west asymmetry, 80
EEE project, 210
Einstein, 53
electromagnetic cascade, 105
Electromagnetic interactions, 39
electromagnetic shower, 152
electron, 8
electron volt, 35
electroscope, 6
electrostatics, 5
energy spectrum, 149, 161
expansion chamber, 40
Explorer 1, 163

Fermi, 128, 183, 249
Fly's Eye II, 171
Fly's Eye, 199
force particles, 33
Fussel, 140

GAMMA Experiment, 195
Gamma radiation, 38
gamma rays, 31, 35, 38
Geiger-Müller counter, 51
geomagnetic cutoff, 69
geomagnetic field, 63
Globus, 222
Gockel, 13
grams per square centimeter, 37
GRAPES-3 Experiment, 196
GZK limit, 189

H.E.S.S. II telescope, 198
Half Value Layer, 36
Halzen, 249
Heitler, 102
Hess, 14
HiSparc, 208

IceCube Neutrino Observatory, 245
Indirect ionizing, 38
inverse square law, 34
ionization, 9
ionization with altitude, 17
isocosms, 78

Jánossy, 142

Kamiokande experiment, 243
Klein, 59
Kolhörster, 17, 22, 56

Lamb, 123
Lemaitre, 78
lepton, 33
Libby, 216
linear attenuation, 36
Lorentz force, 65
Lunar Reconnaissance Orbiter, 166

magnetic lens, 126
magnetic rigidity, 66
Matter particles, 33
McLennan, 11
mean free path, 36
mesotron, 112
Millikan, 24, 26, 48, 62
Millikan Rays, 29
minimum ionizing particles, 42
Muir Lake, 27
muon, 112
Muon Detector Systems, 225
Muon tomography, 226

Neddermeyer and Anderson, 108
neutron moisture sensor, 238
Non-ionizing radiation, 35
NOνA experiment, 253
nucleus, 32

Occhialini, 60, 133
Oppenheimer, 115

Pacini, 18
pair production, 44
Pasteur, 222
Pauli, 91, 240
Pfotzer, 146
photoelectric effect, 43
photographic emulsion, 130
Photons, 43
Pic du Midi, 156
Pierre Auger Observatory, 172, 201
pion, 135, 137
Pions, 152
point counter, 51
positive electron or *positron*, 95
Powell, 132

QuarkNet, 211

Regener, 29
Reines and Cowan, 241
Riggi, 230
Rochester, 155
Roentgen, 9
Rossi, 71, 100, 117, 122, 139, 170
Rossi curve, 101
Rutherford, 9, 33, 90

ScanPyramids Project, 233
Schein, 147
sea level cosmic radiation, 92
Showers of Knowledge, 213
Skobeltzyn, 50, 98
solar wind, 63
Spaceship Earth, 204
Square Kilometer Array, 257
square law, 7
Standard Model, 33
Störmer, 70
Störmer cone, 73
Street and Stevenson, 114
supernova, 182
super-saturation, 40
Swan, 83
Swann, 47

TAIGA Advanced Instrument
  Experiment, 202
Telescope Array Project, 200
Thompson, 8
Tibet AS-gamma experiment, 191
Tinlot, 144

Ultra-high-energy gamma rays, 189

Van Allen, 163
Volcano Ranch, 150, 170
Voyager 1 and *Voyager 2*, 166

Wataghin, 142
Wernher von Braun, 164
Wilson, 11, 40, 41, 50, 58
Wulf, 12

Yukawa, 116
Yukawa and Sakata, 124

Zwicky and Baade, 180

www.ingramcontent.com/pod-product-compliance
Ingram Content Group UK Ltd.
Pitfield, Milton Keynes, MK11 3LW, UK
UKHW021831190726
13853UKWH00003B/1278